THE BATTLE OVER
HETCH HETCHY

THE BATTLE OVER HETCH HETCHY

AMERICA'S MOST CONTROVERSIAL DAM AND THE BIRTH OF MODERN ENVIRONMENTALISM

Robert W. Righter

OXFORD
UNIVERSITY PRESS

Oxford University Press, Inc., publishes works that further
Oxford University's objective of excellence
in research, scholarship, and education.

Oxford New York
Auckland Cape Town Dar es Salaam Hong Kong Karachi
Kuala Lumpur Madrid Melbourne Mexico City Nairobi
New Delhi Shanghai Taipei Toronto

With offices in
Argentina Austria Brazil Chile Czech Republic France Greece
Guatemala Hungary Italy Japan Poland Portugal Singapore
South Korea Switzerland Thailand Turkey Ukraine Vietnam

Published by Oxford University Press, Inc.
198 Madison Avenue, New York, New York 10016

www.oup.com

First issued as an Oxford University Press paperback, 2006

Library of Congress Cataloging-in-Publication Data
Righter, Robert W.
The battle over Hetch Hetchy : America's most controversial dam
and the birth of modern environmentalism /
Robert W. Righter.
p. cm.
Includes bibliographical references and index.
ISBN-13 978-0-19-514947-0 ; 978-0-19-531309-3 (pbk.)
ISBN 0-19-514947-5 ; 0-19-531309-7 (pbk.)
1. Water supply—California—San Francisco Bay Area—History.
2. Water resources development—Political aspects—California—San Francisco Bay Area.
3. Hetch Hetchy Valley (Calif.) I. Title.
TD225.S25R54 2005
363.6′1′0979447—dc22 2004048904

1 3 5 7 9 8 6 4 2

Printed in the United States of America
on acid-free paper

For Benjamin, Dylan, Zachary, and Sarah—
the future

The Hetch Hetchy

HARRIET MONROE

Have you found the happy valley?
  No? then follow—I have seen
    Where it lies.
Shoon and staff—oh, leave your alley!
  Pass the foot-hills, pass the green
    Gates that rise.

Soft it slumbers, locked in granite,
  Cliffs like silver-mailèd knights
    Ranged around
And the mountain breezes fan it,
  Snow-plumed winds from hoary heights
    Glacier-crowned.

There slim waterfalls dash madly,
  Breaking, foaming, thundering
    As they pass
Into blue-eyed brooks that gladly
  Trail their gauzy gowns and ring
    Bells of glass.

There the Rancheria, laughing,
  Down her cleft of granite trips
    Like a girl;
Leaps to meet her lover, quaffing
  Cataracts through foamy lips
    As they whirl.

And Tuolumne the river
  From his plunges mountain-deep
    Rests awhile;
Wind with many a curve and quiver
  Down in flowery glades asleep
    Mile on mile.

Come—'neath plumy cedars lying
  We shall hear his crystal tune
    Filmy soft;
Watch his foamy fringes flying
  Till the mountain-climbing moon
    Rides aloft.

Then the stars will guard our slumbers—
  Never head in royal bed
    Lay so still—
While the stream sings lulling numbers,
  And the ghostly shadows tread
    Where they will.

Oh the golden days that shimmer
  In that deep entrancèd vale
    Richly bright!
Oh the twilights dim and dimmer,
  Till from granite shoulders pale
    Falls the night!

Come, friend, pass the frowning portals!
  Take the Magic Valley—stay—
    It's your quest.
Come, forget that we are mortals—
  Where the gods have had their way
    Men are blest.

FROM

HARRIET MONROE, *YOU AND I*

(NEW YORK: MACMILLAN CO., 1914)

195–7.

ACKNOWLEDGMENTS

IN CONTRAST perhaps to novelists, historians are highly dependent on others. Archivists and librarians deserve special praise. Somewhere in the past they decided to keep material, not knowing whether it might be useful or not. To those past archivists who had the foresight to keep Hetch Hetchy materials, I offer a hearty thanks. Those in the present I can thank directly. The staff at Southern Methodist University were particularly helpful. Michael Foutch helped dig out engineering books and journals, and the staff at Interlibrary Loan assisted in finding some esoteric material. The DeGolyer Research Library held materials on Hetch Hetchy that did, indeed, surprise me. Former director David Farmer, now retired, gave generously of his time. Present director Russell Martin and Betty Friedrich deserve my special thanks. No one can possibly do justice to the Hetch Hetchy controversy without using the invaluable collections of the Bancroft Library, University of California. The staff there, particularly Teresa Salazar, Susan Snyder, Jack Von Euw, and Walter Brem pointed me in directions I might have missed. Also, a special thanks to Jim Snyder, the archivist and librarian at the Yosemite Archives and Library, Yosemite National Park. Archivists at the San Francisco Room of the San Francisco Public Library provided materials unavailable elsewhere. Stanford University; the National Archives, College Park, Maryland; and the Federal Record Center, San Bruno, California, provided interior department and court documents which rounded out the story.

I have, of course, benefited from the help of specialists in the field of environmental history. Michael Cohen, Richard Sellars, and Mark Harvey made many substantive suggestions on one or more of the chapters. I have also profited from conversations with Karen Merrill, Donald Worster, Hal Rothman, Richard Orsi, Richard Lowitt, Richard White, and David Beesley. I owe particular thanks to Dan Flores, who spent hours listening to and commenting on the Hetch Hetchy issue on a memorable backpack trip down the Grand Canyon of the Tuolumne River.

I am appreciative of family members who have taken an interest in my history-writing efforts. My brother, Richard L. Righter, and my son-in-law, Loy Lack, read the entire manuscript and, in stimulating conversation, made numerous suggestions which altered my thinking. Daughters Trisha Lack and Bonnie Sanders, and Bonnie's husband Ron, expressed curiosity and offered shelter on research trips. Father-in-law Atwood Smith kept me working.

The history department at Southern Methodist University has been particularly welcoming to a historian who came along with his wife, Sherry Smith, to Dallas, Texas. As a confirmed Westerner, I find Dallas is a little too far east, but the history department has eased the transition. Ed Countryman liked the Hetch Hetchy idea. Tom Knock helped me to understand the Woodrow Wilson era and Christa Deluzio worked to sensitize me to gender issues. John Mears, with his broad interests in world history, broadened my perspective. David Weber, a fine scholar, showed unfailing interest. I must also thank Jim Hopkins, the chair, for his wonderful encouragement. Other members of the department gave valuable criticism when I delivered a seminar paper to the faculty in the spring of 2002. Finally, the department has been wonderfully generous in providing two grants that helped greatly in travel and in the writing of this book.

Ron Good, director of Restore Hetch Hetchy, generously shared his archival material collection. Patricia Martel, former manager of the San Francisco Municipal Utilities District, was gracious in allowing me a phone interview. Hetch Hetchy photographs enhance this story, and I found many excellent ones at the San Francisco Municipal Railroad headquarters, where Carmen Magana was most helpful, as were Katherine Du Tiel and Joe Cowan. I appreciate their assistance.

I want to also acknowledge my parents, Cornelius and Margaret Righter. They first introduced me to Yosemite Valley, and particularly the high country of Tuolumne Meadows. Furthermore, they chose to live in Northern California, a region where a certain conservation ethic seems part of the soil. Neither are with us today, but I like to think they would be pleased to know of their influence in this work, which springs from the experiences of my early life.

Copy editor Steven Baker deserves my praise and appreciation. Two persons have been involved with this project from its initial stages. Susan Ferber, editor with Oxford University Press, liked the ideas from the beginning and has given unfailing support. She offered extensive and excellent comments on a draft version that really was not ready for her excellent editing skills.

Sherry L. Smith, my wife and fellow historian, spent many long hours with the message of the manuscript and my writing style, always in need of repair and refinement. I would not blame her for the errors of the book, but if there is any graceful writing, she must be given the credit. Above all, she has listened and asked the important questions that are difficult to answer. Why does this story matter? Why would anyone want to read what you have written? I do not know that I have satisfied such questions, but I appreciate that she asks them. I admire her greatly and love her immensely.

CONTENTS

HETCH HETCHY CHRONOLOGY

1858 Spring Valley Water Works is established.

1882 Sending Tuolumne River water to San Francisco by aqueduct and ditch is first proposed.

1892 Sierra Club formed.

1901 Mayor James Phelan files for water rights in the Hetch Hetchy Valley.

1903 Secretary of the Interior Ethan Hitchcock denies San Francisco's permit application for water storage at Hetch Hetchy Valley.

1907 San Francisco rejects the Bay Cities Water Company proposal.

1908 Secretary of the Interior James Garfield grants San Francisco rights to develop Hetch Hetchy Valley.

1909 Establishment of the Society for the Preservation of National Parks.

1910 Secretary of the Interior Richard Ballinger asks San Francisco to "show cause" why the Hetch Hetchy Valley should not be removed from the Garfield grant.

1912 San Francisco hires Michael M. O'Shaughnessy as city engineer.

1912 The Freeman Report is published.

1913 Congress passes the Raker Act, granting San Francisco the right, with many regulations, to build a dam and develop power in the Hetch Hetchy Valley.

1914 Death of John Muir.

1916 Passage of the National Parks Act.

1918 Completion of the Hetch Hetchy Railroad.

1923 Completion of the O'Shaughnessy Dam.

1924 Completion of the Moccasin Creek Power Plant.

1925 Signing of the agreement between San Francisco (SF) and Pacific Gas and Electric (PG&E) to purchase Hetch Hetchy power.

1930 Spring Valley Water Company turned over to the city of San Francisco.

1930 Death of James Phelan.

1934 Death of Michael M. O'Shaughnessy.

1934 Completion and dedication of the Hetch Hetchy system.

1939 Department of the Interior wins its Supreme Court decision that the 1925 SF-PG&E power agreement violated the Raker Act.

1952 SF agrees to sell power to the Modesto and Turlock irrigation districts.

1987 Secretary of the Interior Donald Hodel raises the possibility of removing the O'Shaughnessy Dam and restoring the Hetch Hetchy Valley.

2000 Establishment of the Restore Hetch Hetchy organization.

CAST OF CHARACTERS

Albright, Horace	Opposed the Hetch Hetchy scheme, but worked as a secretary for Franklin Lane. Became assistant director of the National Park Service and then director in 1930.
Badè, William	Religion professor at Pacific School of Religion who was heavily involved in trying to save the valley. After Muir's death Badè edited a number of his works.
Brower, David	Executive director of the Sierra Club and environmental leader who lamented the loss of Hetch Hetchy and supported the idea of restoration.
Bourn, William	Wealthy president of the Spring Valley Water Company.
Carr, Jeanne	Friend of Muir in Wisconsin and during his early years in Yosemite. Married to Ezra Carr and active in furthering Muir's scientific and writing career.
Carr, Ezra	Geology professor at the University of Wisconsin and then the University of California. Influenced Muir's acceptance of the glacial origins of Yosemite.
Colby, William	Secretary of the Sierra Club and an indefatigable worker who understood the political process. He directed the effort to save the Hetch Hetchy Valley.
Cramton, Lewis	Congressman from Michigan who was a friend of the national parks. In the 1920s he exposed

	San Francisco's avoidance of provisions of the Raker Act.
Freeman, John	Nationally known civil engineer who designed the Hetch Hetchy water system.
Giannini, A. P.	Founder of the Bank of Italy, which evolved into the Bank of America. At a crucial point, his bank purchased San Francisco water bonds.
Good, Ron	Executive director of Restore Hetch Hetchy, established in October 2000.
Grunsky, C. E.	San Francisco city engineer appointed by Mayor Phelan in 1900.
Hall, William Hammond	A talented engineer who became involved with the city's efforts to secure water rights on Lake Eleanor and Cherry Creek.
Hearst, William Randolph	Editor and owner of the *San Francisco Examiner*. Strong supporter of the city's desire for Hetch Hetchy.
Hodel, Donald	Secretary of the interior under President Ronald Reagan who first suggested the idea of restoring the valley in 1987.
Ickes, Harold	President Franklin Delano Roosevelt's secretary of the interior who insisted that San Francisco comply with section 6 of the Raker Act.
Johnson, Hiram	Elected governor of California in 1910. Favored municipal ownership of water and power and curbs on the political power of corporations.
Johnson, Robert Underwood	Prominent literary figure and editor of *Century Magazine*. A leader in the fight to save the Hetch Hetchy Valley.
Kent, William	Congressman from Marin County, California. A great admirer of Muir and a California Progressive, he eventually sided with San Francisco.
Lane, Franklin K.	City attorney under Mayor Phelan. Appointed secretary of the interior by Woodrow Wilson. Strong advocate of San Francisco's position.
Long, Percy V.	City attorney for many years, he headed the San Francisco legal team that prepared briefs and

represented San Francisco's interests in numerous hearings.

Manson, Marsdon — City engineer from 1908 to 1912. Although a Sierra Club member, he was an outspoken advocate of San Francisco's position.

Marx, Charles D. — Highly respected civil engineering professor at Stanford. He favored San Francisco's position and often worked toward the city's goal.

Mather, Stephen — A Chicagoan who supported the defender's position. In 1916 he became director of the National Park Service and in the 1920s protected Yosemite National Park from San Francisco's ambitions.

McFarland, J. Horace — Executive director of the American Civic Association, and a consistent defender of the national parks and the Hetch Hetchy Valley.

Monroe, Harriet — Well-known Chicago poet. Founder of *Poetry* magazine. Next to Muir, the most elegant defender of the valley.

Muir, John — Famed naturalist who founded the Sierra Club in 1892. He was the spiritual leader of the effort to save the Hetch Hetchy Valley.

Norris, George — Senator from Nebraska who viewed the Hetch Hetchy fight from the perspective of private versus public electrical power. He favored public ownership.

Olmsted, Frederick Law, Jr. — Son of the famed landscape architect, he strongly supported the preservation of the Hetch Hetchy Valley and would play a major role in passage of the National Parks Act of 1916.

Olney, Warren — Charter member of the Sierra Club. Deeply torn in his views on Hetch Hetchy, he sided with San Francisco.

Parsons, Edward T. — Leader in the Sierra Club. In Muir's later years he assumed more responsibility. Married to Marian Randall Parsons.

Parsons, Marion Randall — An accomplished fiction and nonfiction writer

	who fought hard for Hetch Hetchy and acted as Muir's literary secretary in the last year of his life.
Phelan, James	Prominent, wealthy San Franciscan dedicated to the Hetch Hetchy project. Mayor of San Francisco from 1896 to 1902 and U.S. senator from California from 1915 to 1921. He was deeply involved with the water and power project from 1900 until his death in 1930.
Pinchot, Gifford	Head of the United States Forest Service under Theodore Roosevelt. On a national level, he was the most influential supporter of San Francisco's aspirations.
Raker, John	California congressman who sponsored the Raker Act, giving San Francisco the right to dam the Hetch Hetchy Valley.
Ruef, Abe	Notorious San Francisco city boss. His corrupt dealings in 1906 assured that the city would not consider a viable water alternative to Hetch Hetchy.
Sampson, Alden	Hiker, climber, and friend of Muir. More than once he represented the Sierra Club at congressional hearings in Washington, D.C.
Schussler, Hermann	Highly respected civil engineer for the Spring Valley Water Company.
Watrous, Richard	Horace McFarland's very capable assistant at the American Civic Association.
Wheeler, Benjamin Ide	Influential president of the University of California. He used his friendship with Roosevelt and Pinchot to advance San Francisco's case.
Whitman, Edmund	Boston attorney and prominent member of the Appalachian Mountain Club. He represented the Society for the Protection of National Parks at hearings in Washington.
de Young, Michael	Outspoken editor of the *San Francisco Chronicle.* Strongly favored the city's position.

LIST OF ILLUSTRATIONS

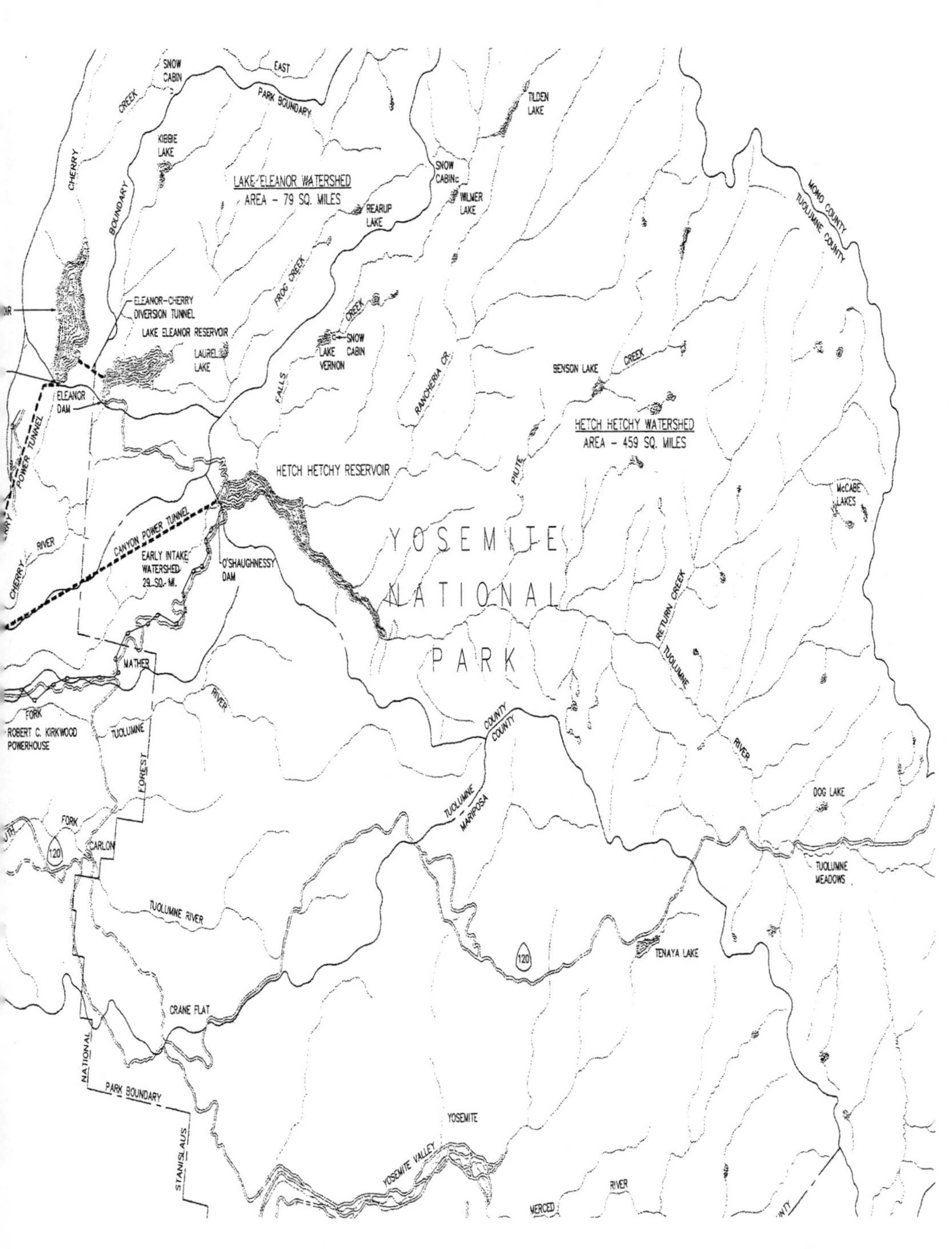

MAP 1. This map shows a modern (1992) view of the mountain section of the Hetch Hetchy water and power system. *Courtesy of the SFPUC.*

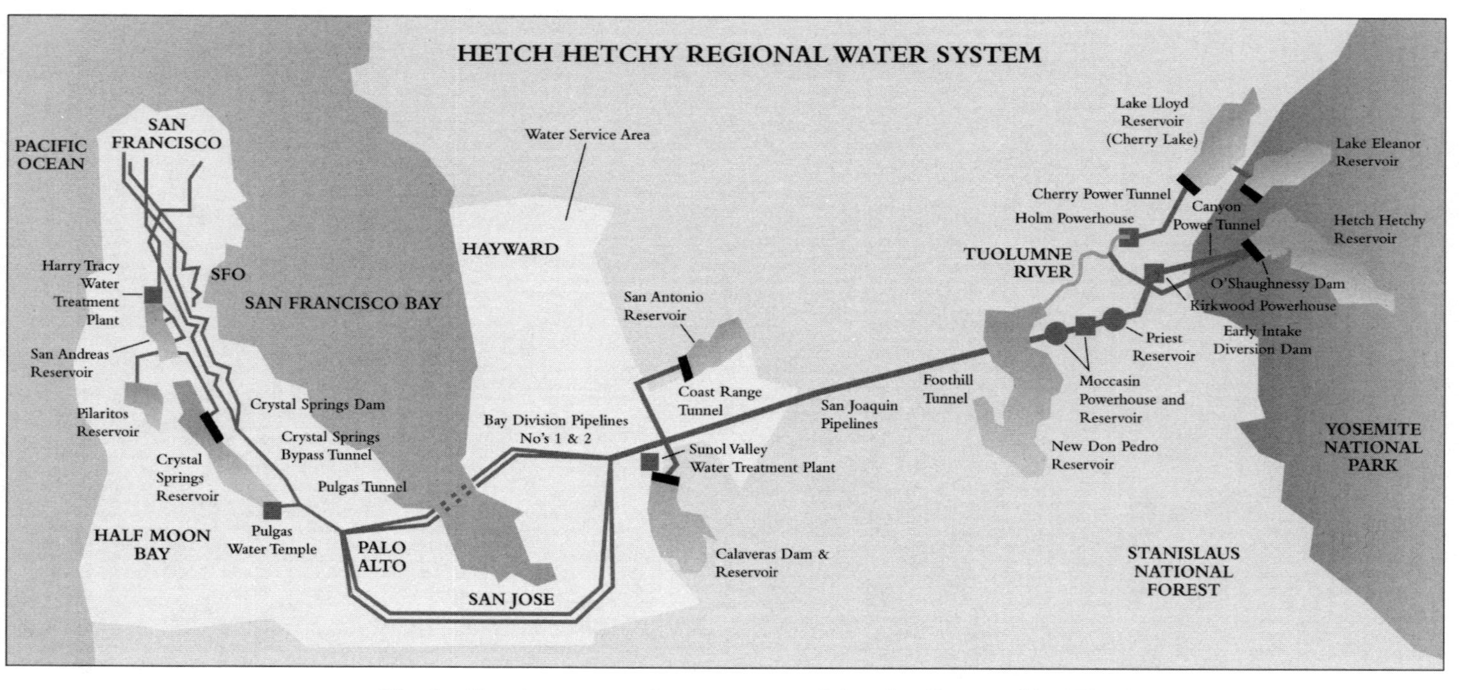

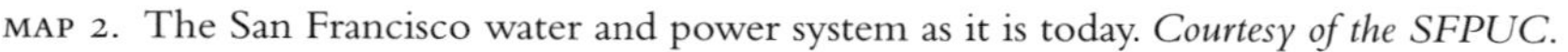
MAP 2. The San Francisco water and power system as it is today. *Courtesy of the SFPUC.*

THE BATTLE OVER
HETCH HETCHY

Introduction

On a grassy knoll overlooking Crystal Springs Reservoir on the San Francisco Peninsula sits an unexpected, Roman Renaissance revival temple. As a child I accompanied my father as he drove past the Pulgas Water Temple to and from our home. Often he would stop at the site to let me stretch my legs and work off a little energy. I would run ahead alongside the reflecting pool, but slow to a walk as I approached the temple. The silence of the pool was replaced by the tumultuous roar of 34 million gallons of Hetch Hetchy water, arriving daily. Hesitatingly, I would climb the steps to peer over the low cement wall and watch the frothing water as its roar echoed and reverberated throughout the temple. The scene was fascinating, yet fearful. My father, a Stanford graduate, told me that freshmen students who joined an eating club had to walk blindfolded on a two-by-twelve-inch board across the abyss as part of their hazing ritual. Whether or not this was true, I resolved that I would never attend Stanford University.

More recently, I revisited the temple and for the first time looked up to read the etched inscription on the cornice: "I give Waters in the Wilderness and Rivers in the Desert to give Drink to my People" (Isaiah 43:20). The verse invoked God's role in a project dedicated in 1934 and completed at the sacrifice of many lives and years of struggle and at an astronomical cost. The temple commemorated San Francisco's achievement in bringing water from the distant mountains to the thriving metropolis. It would serve as a permanent reminder of a remarkable effort, and the notion that the bountiful resources of nature are won, not given.

The achievement, however, was not realized without great controversy. This book is the story of that controversy. The struggle over the fate of California's Hetch Hetchy Valley represents a classic story in the ongoing debate over human land use. It involves water, a valley, and a city. In question was the fate of a spectacular mountain valley located within Yosemite National Park. Would it become a reservoir, storing water for San Francisco's people, or would it remain a natural site, visited by recreationists and developed only to make it more accessible to tourists? The fight, which began in 1900, brought into broad relief conflicting values. It signaled the opening salvo of a century-long conflict over the "highest and best use" of natural areas.

On one side was San Francisco mayor James Phelan, a wealthy, distinguished, ambitious man who understood that his city, surrounded on three sides by salt water, was in trouble. If his dream of a cultured, thriving metropolis was to be realized, he had to find a reliable water supply. Neither groundwater nor nearby surface water was adequate. Phelan and the city leaders looked to the Sierra Nevada mountains, some 150 miles distant. There they found the Hetch Hetchy Valley, a potentially perfect reservoir site with the crystal waters of the Tuolumne River flowing through it. The site, if dammed, would supply the city's water needs for at least 100 years of growth and, hopefully, ensure future prosperity. As an added bonus, the new dam would be the centerpiece of a hydroelectric system to generate power for the metropolitan area.

Those opposed to the mayor's vision understood that the Hetch Hetchy Valley was spectacularly beautiful, nearly as impressive as its sister glacier valley, Yosemite. When leaders of the Sierra Club, formed in 1892, became aware of San Francisco's intentions, they protested. Led by the visionary, eloquent John Muir, the little club mounted determined opposition. Hetch Hetchy lay within the boundaries of Yosemite National Park, established in 1890. Club members and supporters considered the park sacrosanct, protected and free from intrusions. The city's plans represented a deep wedge driven into the integrity of the embryonic national park system. Beyond the park issue, Hetch Hetchy's defenders viewed the waterfalls, granite cliffs, and translucent river as a place of spiritual transcendence, one that should never be desecrated by a dam and a reservoir. In their view the city could obtain its water supply elsewhere.

The Hetch Hetchy fight, at first a regional issue testing competing visions of land use, gained a national spotlight. New priorities emerged, causing some people to question the received wisdom of nineteenth-century land use. For the first time in American history, a national audience listened, and often participated in the debate, as valley protectors questioned water development as the best use of a natural resource. Changing cultural values challenged immediate, material ones. The contenders also debated the dollar value of tourism versus water development, in the present and for the future. They asked difficult questions. Were scenic land and national parks important to the American people? What should be the next human use of the valley? Should it continue to be low impact, preserving the area's integrity, or should Hetch Hetchy be transformed to meet the growing needs of the San Francisco Bay Area? Most Americans knew the answers, but a fresh century ushered in new ideas. For some Americans the idea of progress needed to be redefined, or at least modified. For a small but growing number, scenery and special public lands should be reserved from wholesale development. In their minds the ascendancy of technology and materialism symbolized American civilization gone wrong. It was time to listen to the earnest voices of moderation, rather than to hydraulic engineers with their charts and benefit projections. Five secretaries of the interior, three presidents, and Congress listened, but in the end, legislators passed the Raker Act in 1913, giving the city the right to build the O'Shaughnessy Dam. This resolution, while a victory for San Francisco, has never been fully accepted by the defeated.

Aspects of this story have been told before. When Ray Taylor wrote his book on Hetch Hetchy in 1926, he felt compelled to tell his readers that "more has been written about the Hetch Hetchy than any other subject connected with the city's history. Possibly no subject before a municipality ever was dissected and investigated so thoroughly as the city's water supply."[1] And yet Taylor's book was published eight years *before* San Francisco dedicated the Pulgas Water Temple and Hetch Hetchy water flowed through the city's household taps. Taylor's assessment referred to newsprint articles and official documents weighing in on the case, of which there was no shortage from 1905 to 1913. The issue often found the front page of San Francisco's vigorously competing newspapers, led by William Randolph Hearst's *Examiner,* the *Call,* and Michael de Young's *Chronicle.* Since Taylor's time there have been other books, as well as a number of articles.[2]

A number of issues remain, however. What was this fight all about? We know what San Francisco hoped to accomplish, but what did Hetch Hetchy's defenders want? In effect, if John Muir and the Sierra Club had won the struggle, how would the Hetch Hetchy Valley appear today? In *Wilderness and the American Mind,* Roderick Nash provided the interpretation that every environmental historian has embraced. In an oft-quoted sentence, Nash framed the issue as a battle over the competing claims of wilderness and civilization. I, like many others, accepted this statement unquestioningly, believing that the fight was for wilderness preservation. In the 1960s and 1970s we wanted to believe the Nash thesis. It was so right for the time. The problem with this assertion is that no matter how we might choose to define wilderness, it was not an issue in the Hetch Hetchy fight.

The defenders of the valley consistently advocated development, including roads, hotels, winter sports amenities, and the infrastructure to support legions of visitors. The land use battle joined over one question: Would the valley be used for water storage or nature tourism?

As it turns out, the Hetch Hetchy fight represents the seminal battle not over wilderness, but over public power. San Francisco, indeed California, was in the throes of Progressive-era reform. The Lincoln-Roosevelt League wished to curb corrupt control of city and state government, and with the election of Hiram Johnson as governor in 1910, the reformers positioned themselves to pass needed legislation. On a national stage, Hetch Hetchy became caught in the cross fire between the interests of private utilities ownership and those of municipal ownership. Put another way, if Congress denied the city of San Francisco the Hetch Hetchy Valley, the California Progressive leaders suspected that it would only be a matter of time before the emerging Pacific Gas and Electric Company would grab the area. If, on the other hand, San Francisco gained control, it would signal an important victory for public power, resulting in lower rates for the people. Senator George Norris of Nebraska cared little about water use or recreation potential but considered public ownership of the generation and distribution of electricity essential. He, as well as a number of legislators, believed that Muir and his friends were mere puppets, duped by Pacific Gas and Electric and the Spring Valley Water Company to do their bidding. The defenders of the valley never escaped the charge of collusion with private water and power interests, and it was a major reason why they lost.

The story of Hetch Hetchy did not end, of course, with the passage of the 1913 Raker Act. An ongoing battle over the conditions of the act flowed through the current of the twentieth century. Often city leaders, frustrated by the Raker Act's regulations, wished that they had never heard of Hetch Hetchy and the Tuolumne River. In retrospect, leaders could have met the city's water and power needs in less perplexing and less expensive ways if they had chosen one of the alternative sites.

Inappropriate engineering decisions head the list of problems. Although John Freeman, the engineer who designed the Hetch Hetchy system, commanded the respect of his profession, he was a man who liked to think big—too big. His system produced far more water than the city could consume, and cost far more money than the city could afford. Freeman believed in what we call today "economies of scale": build big and let the city grow into its future projected population. This may be wise planning, if you can bear the expense. However, with no federal or regional funding, the city depended upon its citizenry. Could the city afford it? Initially the voters said yes, but then they said no. The result was that the city received only 50 percent of what Congress granted under the Raker Act. In other words, Congress anticipated that San Francisco would establish municipally owned water and electrical systems. With great difficulty, the city finally purchased the water system, but the idea of public power soon became a distant memory. Today the Pacific Gas and Electric Company still owns the city infrastructure and provides electrical power to companies and residents. The Hetch Hetchy colossus, with its insatiable appetite for dollars, partially explains the city's rejection of public power. An engineering design that could be adjusted to the needs and growth of the city would have made better sense.

Because Freeman designed such an elaborate system, the challenge of its construction comprises a story in and of itself. Civil engineer Michael O'Shaughnessy took on the challenge of building a mighty dam high in the mountains, transporting pure water some 167 miles, and generating electricity along the way. Aside from the Panama Canal, and perhaps New York City's Catskill Mountains water project, it was the largest civil engineering undertaking of its day. In this demanding enterprise engineers faced problems and devised new solutions. Many hundreds of workers labored in a difficult environment. They took pride in their accomplishment, and I have tried to tell a little of their long-neglected story.

However, the greatest significance of the Hetch Hetchy fight lies in the legacies it has left us. Past decisions often have unforeseen consequences. Gifford Pinchot, the head of the United States Forest Service and a stalwart supporter of San Francisco's claim for the valley, devoutly wished to bring the national parks under the mantle of his agency. However, by his vigorous support of the city's position, he alienated influential people who would largely determine the future leadership of the parks through the National Parks Act of 1916. In effect, he ensured that his desire would never be realized. In a classic sense, he won his battle but lost the war. Moreover, ever since workers cleared the valley and built the dam, the Hetch Hetchy story has been involved or invoked in many fights over land use, often causing bureaucrats or civil engineers to wince or squirm at mention of the famed controversy. As an example of what should *not* be done to a scenic mountain valley, the story has no equal. While the sacrifice of Hetch Hetchy is not the sole factor, most dam proposals in national parks over the course of the twentieth century have been contested and successfully stopped.

Part of the legacy of Hetch Hetchy has surely been in the realm of historical memory. Many people, particularly Californians, are aware of the controversy. It is a saga that people with claims to environmental sensitivity can recite. Because it is the keystone to the San Francisco Bay Area water system, it has strong name recognition. Area residents, who consider Hetch Hetchy an ongoing issue, raised a furor when Secretary of the Interior Donald Hodel proposed in 1987 that the O'Shaughnessy Dam might be breeched and the valley restored. The book's final chapter suggests that the reservoir may not be the final chapter in the story of the submerged valley. As one advocate of restoration told me in a hushed voice, "The valley is holding its breath." In 1913 San Francisco and its allies prevailed, but the final step, and one increasingly advocated by environmentalists, might be the breaching and removal of the dam and the restoration of the flora and fauna of the valley. Such a proposal is fraught with both financial and resource concerns. Although San Francisco officials have shown little interest, they may be in denial. In the new century, engineers will likely dismantle more dams than they build. In the years to come we can expect more proponents of restoration to arise.

Should San Francisco have fought for possession of the Hetch Hetchy Valley, and then made it, once acquired, the centerpiece of the city's municipal water system? For many, the answer is clearly no. The beauty of the valley is certainly a factor. But more important, alternatives existed. San Francisco officials became obsessed with developing Hetch Hetchy, to the point that they dismissed or discarded other rivers and valleys that would have served them better. Engineers became enchanted with the valley's water-storing capacity, as if it was created for their purpose. They were also fascinated with the daring challenges of creating a gravity water system originating high in the mountains and building a lengthy aqueduct to get the "product to market." Politicians fell under the sway of the engineers. They depended too heavily on the authority of technical experts. Rather than question the objectivity of city engineers Marsden Manson and Michael O'Shaughnessy, city politicians deferred to their judgment. Both men, competent in their profession, possessed the unfortunate trait of inflexibility.

The city turned its back on at least three respectable water alternatives to Hetch Hetchy. Regrettably, men of questionable character presented two of these options, thus precluding serious consideration. Later Oakland and the East Bay cities took up one of the alternatives—the Mokelumne River—creating a pure water system in less time, at much less expense, and with little political discord and no environmental controversy. Downriver from Hetch Hetchy the Modesto and Turlock irrigation districts built the Don Pedro Dam, and the enlarged structure today stores, mainly for irrigation purposes, almost six times the amount of water as the Hetch Hetchy Reservoir. There were plenty of viable alternatives, and the defenders of the valley pointed them out. Stubbornness, political power, and a sense of entitlement sent San Francisco leaders down a costly path. Initially opportunities to reconsider the city's direction occurred, but as both money and egos became invested, side trails became difficult and turning back, impossible. By 1930 San Franciscans could only shake their collective heads in wonderment, and continue building and spending money to bring the project to completion. Today, city leaders praise the system. Hetch Hetchy delivers abundant and pure water, as well as valuable electrical power. However, to simply say that the end justifies the means is to deny the importance of reflection and of history. Flexibility and resilience among San Francisco's leaders could have resulted in an equally fine water system, built at less expense, without the invasion of a national park and the loss of a remarkable valley.

THIS BOOK reflects more than an academic interest. Since my childhood experience with the Pulgas Water Temple, I have been linked to Hetch Hetchy. My parents raised me in the San Francisco Bay Area. Settling in the "Peninsula" community of Burlingame meant we consumed Hetch Hetchy water and seemed to be quite healthy for the experience. But for a youth, water from a spigot or faucet held no particular attraction. More intriguing was the Crystal Springs Reservoir and the extensive San Francisco watershed. The reservoir stored Hetch Hetchy water after its long journey from the Sierra Nevada mountains. A short hike from my home, the reservoir and watershed lands were forbidden land, de facto wilderness and were closed to all human trespass, save officials and employees of the San Francisco Public Utilities Commission. That, of course, was reason enough for an adventurous 12-year-old to climb the fence and enter what was off-limits. Twice a friend and I hiked over the hills to Lake Pilarcitos. The steep, wooded, brushy, trail-free land was impenetrable, so we cautiously kept to the gravel road, quick to scurry into the brush should we hear an official car or truck. But there were few cars, and in many ways the San Francisco Public Utilities Commission offered me my first experience with a natural environment.

In the summer, my parents took my brother and me camping in the Sierra Nevada. On our way we often stopped at the John Willms Ranch, near Knight's Ferry on the Stanislaus River. My mother's grandfather founded the ranch. He and a friend homesteaded in the late 1850s and, by adroit use of land laws, built up a sizable land base. My family still runs part of the ranch, and often my parents stopped to see relatives or check on the condition of the grass. Across the property curve three, mainly buried, 48-inch pipes—the Hetch Hetchy aqueduct. The ranch itself is sometimes the direct recipient of Hetch Hetchy water when the pressures of the gravity system back water out onto a neighbor's land and the spillover flows down to what I call our Hetch Hetchy pond.

My parents' mountain destination was Tuolumne Meadows. Sometimes we camped in tents, but usually Mom and Dad pulled their 16-foot trailer up the Big Oak Flat Road and then negotiated the Tioga Pass Road's 21-mile horrendous stretch. Once in the meadows, our trailer would stay for the summer. Obviously, it was a different era, one in which National Park Service rangers knew the regulars, and worried about neither camping time limitations nor fees. We would come and go, and in a sense it was like having a small cabin in the heart of the Yosemite high country. By the time I was fifteen, the small trailer became a base camp from which I would take day hikes and increasingly long backpack trips. At age 17 I hiked 150 miles of the John Muir Trail with 2 friends.

John Muir, of course, is the touchstone of the whole Hetch Hetchy struggle. I knew of Muir because I was raised in Northern California. No doubt my parents spoke of him, although they were not members of the Sierra Club nor were they inclined to rhapsodize over nature or wilderness. My appreciation of Muir emerged from my connection with his country—the Yosemite. After college I retreated to the trails of Yosemite. I worked at various jobs in Yosemite Valley and later spent two summers as a Tuolumne Meadows Lodge employee. I devoured Linnie Marsh Wolfe's Pulitzer prize–winning book, *Son of the Wilderness: The Life of John Muir,* shortly after it was published. Wolfe's wonderful portrait of Muir and my own association with his country made me one of Muir's large cadre of admirers. For my part, I remained unsure if Jesus Christ was the son of God, but I knew that John Muir was. I was in awe, like William Keith, his friend and artist. Keith met Muir in Yosemite Valley in 1872. The wild Muir emerged from the high country, Keith recalled. "We almost thought he was Jesus Christ. We fairly worshiped him!"[3] His end-of-life struggle to save a valley that few had ever seen suggested to me, at the time, Muir's cross of Calvary borne for Nature. I have since realized, as did Keith, that my ideas about Muir were a bit romantic. As this book suggests, he was not divine, but rather very human. He did not die from the loss of Hetch Hetchy, but rather from natural causes. On the other side, the San Franciscans were not from the kingdom of Hades, but rather honorable men committed to doing what they believed was right.

For many years Hetch Hetchy has absorbed my interest, both as a place and a subject. I can only hope that this work creates a fresh, more complete, more nuanced understanding of its story, one that will continue to reverberate from California out across the nation.

CHAPTER 1

The Uses of the Valley

"We lay by the fire and revealed our inmost selves . . . until we were overcome by sleep."

ROBERT UNDERWOOD JOHNSON

"IMAGINE YOURSELF in Hetch Hetchy," wrote John Muir. "It is a bright day in June; the air is drowsy with flies; the pines sway dreamily, and you are sunk, shoulder-deep, in grasses and flowers."[1] These are the words Muir penned to describe his first visit to the enchanted valley. Alone, he sauntered up and down, east and west, describing cliffs, waterfalls, and various wonders. From this 1871 exploration he seemed to bond with the valley, and his rhapsodic prose reflected his infatuation. Freed from the tame, rolling hills of his Wisconsin youth, Muir exulted in vertical landscape, in high mountain meadows, in granite, in ancient redwoods, and the exuberance of sparkling waterfalls. Yosemite Valley often claimed his attention and passion, but the Hetch Hetchy Valley, a geological replica, did not suffer Muir's neglect. The valley represented all that he loved, and it would never be far from his thoughts until his death in 1914.

Muir delighted not only in Hetch Hetchy's grandeur but also in its geology, and this dual interest reminds us that humans used the Hetch Hetchy Valley in different ways. The Central Miwok and Paiute Indians tribes certainly altered the landscape, while European American miners fervently wished to do so on a larger scale. Sheepherders with their flocks changed the nature of the valley grasses and ferns. Writers and artists, such as Muir and Albert Bierstadt, recorded their impressions, but their enthusiasm also represented change through visitation. The nineteenth-century story of Hetch

Hetchy was, then, one of human agency, a progression of use in which people with varied interests gently used the valley, but without compromising its material wholeness or its integrity.

Muir's enthusiastic, gazelle-like rambles throughout the Yosemite high country were surely motivated by sheer pleasure, but also by a scientific purpose. He was convinced of the glacial origins of the domes and canyons of Yosemite Valley, but a "higher authority" thought otherwise. Josiah Whitney, Harvard professor and head of the California Geological Survey, had pronounced that Yosemite Valley resulted from a cataclysmic event in which the valley subsided when "support" was "withdrawn underneath." With unjustified scientific certainty Whitney pronounced that the glacial origin idea represented an "absurd theory . . . based on entire ignorance of the whole subject."[2] To contest Whitney's beliefs, Muir observed not only Yosemite Valley, but also what he would often call his "Tuolumne Yosemite." There

FIGURE 1. This is the first photograph taken of the valley, probably by a member of the Whitney Survey in 1869. *Courtesy of the Bancroft Library.*

he studied the "Great Cañon of the Tuolumne," envisioning an immense ice river scouring the canyon, creating a chain of lakes that would eventually be filled with glacial drift. Hetch Hetchy Valley evolved from one of those ancient lakes. The similarities of Hetch Hetchy with Yosemite Valley were definitive evidence for the young Muir. He concluded that "the Yosemite Valley is a cañon of exactly the same origin" as Hetch Hetchy.[3] He disputed Whitney's description of the former as an "exceptional creation." In a philosophical mood, he noted that "among the endless variety of natural forms, not one stands solitary and unrelated."[4] Thus Hetch Hetchy would always be especially close to him because, along with mentors such as geologists Louis Agassiz and Ezra Carr, it was a great teacher and Muir was a devoted student.

This spectacular geology fated the Hetch Hetchy Valley for controversy. Those who initiated the dispute also found the valley attractive, but for very different reasons. One such man was San Franciscan James Duval Phelan, born of wealth, a native Californian destined to make a difference. While Muir worshiped his valley, Phelan loved his peninsula. At the tip of this peninsula, like a fingernail, spread the city of San Francisco, a noble city that looked to the Pacific Ocean and Asia on its west and to an inland empire on its east. Phelan hoped to forge San Francisco into an imperial center, stretching its economic tentacles in all directions. His vision extended beyond mere wealth to include building a "city beautiful," featuring wide boulevards, distinguished architecture, classic public buildings, grand fountains, parks, vistas of the surrounding water, and of course, social harmony.[5] But location posed a problem. Water surrounded the city, but salt water would slake no one's thirst. One can imagine Phelan, with his penchant for poetry, looking out from the hills of his city and reflecting on Samuel Coleridge's "Rime of the Ancient Mariner":

> Water, water, every where,
> And all the boards did shrink;
> Water, water, every where
> Nor any drop to drink.[6]

To realize Phelan's imperial vision, the city had to have fresh water in large quantities. So he looked to the pure streams of the distant mountains and he dreamed of the aqueducts of Rome. Although other Sierra water sources could be tapped, Phelan wanted the Hetch Hetchy Valley, and as his friend novelist Gertrude Atherton noted, "he could talk the hind legs off a donkey, and when he applied himself to win a point he won it."[7] For Muir, Phelan would prove a determined adversary.

Phelan's desire to harness the valley represented the final phase of human use. Long before, perhaps 3,500 years earlier, Native Americans found Hetch Hetchy attractive for very different reasons. These first humans relished the valley as an escape from the summer heat of the San Joaquin Valley, and perhaps as well from the winter cold. The valley is only 3,800 feet in elevation, a paradise for a Paiute Indian from the east who had labored up Bloody Canyon and over frigid 10,000-foot Mono Pass. But daily subsistence occupied much of California Indians' lives, and the valley provided for their needs through the harvesting of black oak acorns and the seeds of grasses. Fishing in the river and hunting in the vast meadows and nearby mountains were presumably common activities. Evidence suggests that these people adjusted to their impressive environment, allowing them to live well.[8]

It would have surprised Muir to learn that the Hetch Hetchy Valley he enjoyed in 1871 was not really a wilderness, but rather a landscape managed by California Indians long before he or any other Euro-American ever laid eyes on it. It was their home, at least in the summer and fall. The Central Miwok and their ancestors had inhabited the valley as long as they could remember. They did not change it drastically, but they did alter it to their liking. Of course, the great granite cliffs and waterfalls could not be shaped by human hands, but the valley could. Anthropologists know that the California Indians used fire often in the western foothills of the Sierra Nevada and in such valleys as Yosemite and Hetch Hetchy.[9] Burning would increase the grasses and ferns, which would, in turn, increase the population of deer and other game. It would also clear away brush, making the narrow valley more open and easily traversed. It is, of course, impossible to compare the Hetch Hetchy Valley of 1870 with the valley today, but an 1866 photograph of Yosemite Valley taken by Carleton Watkins reveals much more meadow and open space than a rephotograph taken in 1961, the difference likely the result of recurrent fire, possibly natural but probably man-made.[10]

Aside from Hetch Hetchy's abundant meadows, its great attractions were the black oak trees (Muir called them Kellogg oaks and they are given the Latin name *Queercus kelloggi*), which provided valley Indians with bumper crops of acorns that they annually stored in what they called "chuck-ahs." Compared to the canyon live oak, or scrub oak—a tree one sees today on the trails surrounding Hetch Hetchy—the black oak was more massive, a substantial tree in both productivity and longevity. These mountain dwellers, according to anthropologist Helen McCarthy, were the Indians' "preferred species in many localities." Because the black oak was such a valued food source, we can assume that if natural fire (lightning) did not clear the competing brush, the Indians would have. Although they valued these Hetch Hetchy oaks, or

FIGURE 2. This small but noisy waterfall marked the entrance of the Tuolumne River into the Hetch Hetchy Valley. *Courtesy of the Bancroft Library.*

"acorn orchards," as Muir called them, the California Indians did not purposefully plant or tend them. It would take nearly 30 years for the first acorns to appear on a black oak. At 80 years one could expect a bumper crop, and at 175 years the black oak was fully mature. One Mono Indian laughingly remarked that she would never plant a black oak acorn: "I'd be dead before it was grown. Let the blue jays do the planting, it's their job."[11]

In harvesting the black oak acorns, the Hetch Hetchy Valley Indians may have used a technique called "knocking."[12] Using long branches or poles, Indians would hit or shake the branches of the oak, thus loosening the acorns. In this way they might win out over competitors such as blue jays, woodpeckers, squirrels, gophers, deer, and even bears, of which there were many in the Hetch Hetchy Valley. Such a technique would also prune the trees of dead branches and encourage lateral growth. The downside was that the black oak acorns took two years to mature, so knocking the outside branches might result in the loss of some of next year's crop. Nevertheless, that trade-off might have seemed worthwhile to the Indians, especially considering the large population of black bears and, no doubt, some grizzlies. Thus it was that these Central Miwok bands knew the Hetch Hetchy Valley as a welcoming place in the summer, a secluded valley where they might hunt but would primarily gather and store the treasured black oak acorns. To enhance both these activities, they, like all human beings, altered their environment to make their lives easier.

It was likely either the Ahwahneechee or the Tuolumne tribe—both bands of the Central Miwoks—that gave the valley its peculiar but memorable name. *Hetch Hetchy* is thought to be a derivation of the word *hatchatchie,* which refers to a species of grass with edible seeds that once grew in the lower end of the meadow. In late summer, when the seeds had ripened, Indians could be found gathering the grain and then pounding it into meal for porridge.[13] One early white visitor noted that the natives were preparing a variety of grass and edible seeds. When he asked the name, they responded "hatch hatchy," providing the valley's nomenclature.[14]

However, another account disputes such an interpretation. According to Tenaya, an Ahwahneechee chief associated with Yosemite Valley, *hetchy* means "tree." Supposedly at the end of the Hetch Hetchy Valley, where a trail entered the meadow, two yellow pine trees thrived; hence, "Hetch Hetchy," or the "Valley of Two Trees." We will likely never know the authoritative name origin, which perhaps allowed William Jennings Bryan, caught between the passions of Muir's and Phelan's view, to quip that he had been led to believe that Hetch Hetchy was some sort of Indian war dance.[15]

Precision regarding the first Euro-Americans' visit seems equally elusive. Spaniards limited their exploration to the coastal regions of California; thus it was near 1850 that one or more Americans penetrated the Hetch Hetchy Valley, seeking placer gold and, if that was unavailable, at least some game to hunt.[16] Food was scarce and expensive, and gold was increasingly difficult to find; so the "forty-niners" began to reconnoiter the rivers and valleys to the east. With their insatiable optimism and energy, they explored the headwaters of every river and stream in search of "color" and, perhaps, even the "mother lode." In all likelihood, miners stumbled on the Hetch Hetchy Valley as they struggled up the lower canyon of the Tuolumne River in a search that left no stream untried. Ironically, the first white to enter Hetch Hetchy may not have been an American but rather Jean-Nicolas Perlot, an adventurous Belgian miner who spent a great deal of time in the Sierra Nevada foothills. He, a fellow miner named Debrai, and his dog, Miraud, explored up the branches of the Tuolumne River, and according to his account, he may have spent a night in the valley.[17]

We can be a little more certain regarding the Screech brothers. Joseph and Nathan spent plenty of time in the mountains, hunting and prospecting. On one outing for deer or bear, Nate spotted a deep valley to the east, but it was too far to visit. Nate later recalled that "on getting home I asked the Indian chief the name of the valley and he said . . . 'there is no valley. It is only a cut in the hills through which the Tuolumne River runs but if you think there might be a valley keep looking and if you find such a place I will give it to you.'" Nate went on looking for the valley, and in a couple of years he entered the western opening and then "walked up toward the center and faced the Indian chief and his wives." The chief instructed the women to pack up and prepare to leave. He intended to keep his promise, saying to Nate, "The valley is yours."[18] We are obviously treading on the path of mythology here, but the Screeches remain the first non-Indian visitors of record. John Muir designated Joseph as the "discoverer" who laid out the first trail and claimed the valley for his use. Although the Indian-chief-and-gift story is unlikely, the brothers undoubtedly encountered Indians living in their bark shelters, collecting acorns, and living peaceably. As time passed, more Americans arrived, finding remains of Indian occupation but no Indians. In all likelihood the gold rush activity from 1850 to 1855 forced out the Central Miwoks, who moved to Yosemite Valley to work in tourist-related fields or to labor in the mines. The miners were not known for their compassion, and they probably cleared the valley, forcing the displaced Indians to find a home elsewhere.

These early miners found no gold along the banks of the Tuolumne, which surely saved the river and valley from desecration. Since they were itinerant and sometimes illiterate, they left no written record. Their thoughts as well as their eyes were cast downward in search of gold-bearing gravel. Miners seldom could be enticed from their purpose by sheer cliffs or imposing waterfalls. These early Argonauts did, however, note the luxuriant meadows with their profusion of ferns, flowers, and grasses. Within a short time a few of these hopeful Americans, discouraged and often destitute, turned to more prosaic livelihoods, such as sheep raising. They remembered the Hetch Hetchy Valley. By 1860 wool growers instructed their drovers to seek out the level, fertile pastures. The 1,200-acre meadow provided nourishment for "thousands of head of sheep and cattle that entered lean and lank in the spring, but left rolling fat and hardly able to negotiate the precipitous and difficult defiles out of the mountains in the fall."[19] This account exaggerates in number, but as grazing competition increased, wool growers and cattlemen filed homestead land entries to protect their rights. They put up rude cabins, intended for summer use only. When Muir descended into the valley in 1871, he explored "a couple of shepherd's cabins" as well as some Indian huts, which he thought rather quaint but not intrusive.[20] It was late in the season, early November, and the "hooved locust," as he delighted in calling sheep, had long departed for lower elevations.[21]

There were other early visitors. Perhaps the most famous was Josiah Whitney. The professor wrote a glowing description, characterizing the valley as "almost an exact counterpart of the Yosemite." While denying Yosemite Valley the possibility of a glacial origin, he accepted that the "beautifully polished" rocks of Hetch Hetchy were surely evidence of slow, powerful glacial action. For Whitney, as well as Muir, the valley was useful to explain basic origins of land masses. In his opinion "if there was no Yosemite, the Hetch Hetchy would be fairly entitled to a world-wide fame." He encouraged visitation "if it be only to see how curiously nature has repeated herself." In describing the three-mile-long valley, Whitney noted that "a spur of granite" (Kolana) split the vast meadow into two parts. The waterfalls also drew his attention, and he exclaimed that "Hetch Hetchy Fall" (Wapama) was so large in the spring that "the whole of the lower part of the Valley is said to be filled with spray." This lower, most westerly meadow terminated in "an extremely narrow canon, through which the river has not sufficient room to flow at the time of the spring freshets, so that the Valley is then inundated, giving rise to a fine lake." It was, of course, this narrow outlet canyon that would attract San Francisco engineers, although they christened Whitney's "fine lake" a mosquito-filled swamp. But Whitney saw only the positive and the aesthetic, describing the 1,800-foot granite cliffs and the waterfalls with surprising passion, noting that "were they anywhere else than in California, they would be considered . . . wonderfully grand."[22] Whitney's topographer, Charles F. Hoffman, visited the valley shortly after, reiterating his chief's enthusiasm.

But few Californians read either Whitney's or Hoffman's account. Those who did cared much more for information regarding mineral wealth than for scenery. John Muir's 1873 article in the *Overland Monthly* opened a few more eyes to the valley's attractions, but generally it was unknown to all but a handful. This obscurity distressed Muir, and he noted that if he could place the 10,000 Americans who had visited Yosemite Valley by 1873 in the Hetch Hetchy Valley, they would not know the difference. Their only questions might be "What part of the Valley is this? Where are the hotels?"[23]

Although Muir urged his readers to visit Hetch Hetchy, few did. Travel was difficult. There were no wagon roads and no facilities. Commercial interests did nothing to promote the valley. Guidebooks to the Yosemite country generally mentioned Hetch Hetchy only in passing or not at all. The best known of Sierra Nevada boosters, James M. Hutchings, totally ignored Hetch Hetchy in his guidebook, while devoting some five hundred pages to Yosemite Valley. Of course, Hutchings ran a tourist operation in Yosemite and thus saw no advantage in publicizing a competing landscape.[24]

While most early photographers and artists followed Hutchings's advice and ended up working their magic in Yosemite Valley, Hetch Hetchy attracted at least one famed landscape artist. Albert Bierstadt came to San Francisco with his paints and palette in the fall of 1872, looking for sublime mountain scenery, the kind that would guarantee him an income as well as enhance his reputation as a landscape artist. He visited Yosemite Valley, of course, but an encounter with Jeanne Carr, wife of University of California geologist Ezra Carr, led him to the lesser-known valley.[25] Carr suggested that if Bierstadt visited Hetch Hetchy, he would not be merely replicating other artist's impression of Yosemite Valley, but rather creating images of a hitherto unobserved, but equally sublime, mountain valley.

Bierstadt found the idea intriguing, and in July 1873 he set out with his wife and three friends on a six-week pack trip to the Hetch Hetchy region. We do not know if he consulted with Muir, but the artist found the valley as enchanting as the writer. He noted the inaccessibility of Hetch Hetchy but told friends it was worth the effort. "It is smaller than the more famous valley," he explained, "but it presents many of the same features in its scenery and is quite as beautiful."[26] When Bierstadt emerged from the mountains in late August, a San Francisco reporter noted that he was "laden with sketches, from a spot which as yet is almost untrodden soil." We know that from those sketches Bierstadt produced at least four oil paintings, possibly more.[27] In 1876 Mount Holyoke College acquired one of Bierstadt's paintings—simply called *Hetch Hetchy Canyon*—to hang in its art museum, an example of the impact of art in publicizing the little-known valley.[28]

Other artists, scientists, and adventurers trickled into the valley toward the close of the century, often at Muir's urging. That was the case with Charles Dormon Robinson. In the early 1880s the artist had risen in prominence as a landscape specialist, challenging Thomas Hill as the dean of Yosemite painters. Looking for new subjects, in 1884 Robinson ventured on an overland trek to Hetch Hetchy, recording his impressions and even writing a description of the trip for the *San Francisco Chronicle*. More literal in his style than Bierstadt, Robinson's depictions give us an accurate view of the valley.[29]

One other artist who found inspiration in Hetch Hetchy was William Keith, Muir's Scottish friend. The two often traveled together in the Sierra Nevada, and Keith recorded that in 1873 he and Muir undertook a life-

FIGURE 3. "Little Joe" LeConte took a number of superb photographs shortly before loggers and workers stripped the valley. This scene is of the lower meadow with Wapama Falls in the background. *Courtesy of the Bancroft Library.*

threatening descent into the Tuolumne Canyon, the first "white party" to do so.[30] No paintings resulted from that difficult trip, but in the years to follow, Keith recorded his mountain impressions. He spent two memorable Hetch Hetchy outings with Muir in 1895 and 1907. Unfortunately, the earthquake and fire of 1906 destroyed some of his earliest and finest works. Keith and other artists gave much needed publicity to the valley, a place far from the amenities of civilization. Muir occasionally felt compelled to comment on Keith's visual interpretation of the valley. Once, in Keith's studio, Muir complained that a Hetch Hetchy waterfall (probably Wapama) was too thin. Any man can see, insisted Muir, "that the great masses of snow in the vast watershed that leads down to the brink of that waterfall would result in a thundering cataract and not a mere thread like that." Keith feigned outrage, but then, at Muir's urging, he brushed in a more ample waterfall.[31]

Besides promoting the seldom visited valley, Keith and the other artists, and a few photographers as well, were engaged in more than recording special landscapes and making a living. They were explorers in the sense that they were uncovering the aesthetic wonders and visual frontiers of the West. Just as the forty-niners sought nuggets of gold, these painters sought nuggets of Nature. They provided the American people with a new perspective and appreciation of the American West. Such artists gave the nation a sense of pride in the uniqueness of the land that only a few had or would know firsthand. It has been said that European nations gain their identity from antiquity—from civilizations that preceded them. Americans, however, would find meaning in the natural spectacle, much of it found in places like the Hetch Hetchy Valley. To some degree it became a national—or natural—icon.

While Hetch Hetchy Valley languished in isolation, Yosemite Valley was attracting worldwide attention. All of this notoriety resulted in the first step in the establishment of Yosemite National Park, of which Hetch Hetchy Valley would become a part. Yosemite Valley had been in the possession of Indian tribes, primarily the Ahwahneechee, who lived a life of hunting and gathering, moving in and out of the valley—and altering it—at will. That lifestyle changed in 1851, when the almost inevitable clash between American miners and local Indians led James Savage and his militia, known as the Mariposa Battalion, to mount a reprisal expedition. Up the Merced River they went, intent on capturing Chief Tenaya and his band. In the course of that pursuit Savage "discovered" Yosemite Valley. Tenaya escaped, fleeing across the mountains to join the Mono Indians. In time, the Central Miwok Indians returned to Yosemite to take up a rather dual existence that one visitor described as a "curious mixture of tradition and civilization." While still gathering acorns and storing them in their "chuck-ahs," family members might earn wages by washing and ironing white men's clothes, while others would industriously weave baskets to sell to tourists.[32]

While herders drove thousands of sheep into Hetch Hetchy in the late 1850s, Yosemite Valley played host to increasing numbers of curious Americans. With the boosterism of the young Englishman James M. Hutchings, Yosemite Valley moved from obscurity to renown within a decade. With their talent for description, such visitors as Horace Greeley, the Reverend Thomas Starr King, and geologist Clarence King spread the word. The creative ability of photographer Carleton Watkins and artists Thomas Ayres and Bierstadt augmented immeasurably the writers' efforts. By 1863 the famed landscape architect Frederick Law Olmsted, particularly concerned with the American instinct to "privatize" anything of scenic value, was busy advocating Yosemite Valley as a public park.

It was, no doubt, Olmsted who led a group of Californians "of fortune, of taste and of refinement" to Senator John Conness's office in early 1864 to ask that he introduce a bill granting custodianship of Yosemite Valley to the state of California. The bill required the state to agree to hold the valley "for public use, resort and recreation for all time."[33] California, therefore, was not free to do with the valley what it might, but rather became its trustee for the federal government. When a preoccupied President Abraham Lincoln signed the bill into law on June 30, 1864, he authorized a state park with certain federal conditions, but the new law did prove to be a model in concept and language for the Yellowstone National Park Act of 1872.

While the Yosemite Park Act of 1864 provided a modicum of protection for the Yosemite Valley and a small grove of *Sequoia gigantea* (the Mariposa Big Trees), it had no effect on the valley 20 miles to the north. As public domain, Hetch Hetchy was open to those who wished to visit it for pleasure or claim it for economic gain.[34] Such was the status of this secluded place until 1890.

The sheep-grazing free-for-all in Hetch Hetchy and elsewhere threatened to denude the High Sierra meadows, as did lumbermen in the nearby forests. The threats drove Muir into action. The hundreds of thousands of high-country acres surrounding Yosemite Valley needed protection through permanent park status. That certainly was high on the agenda when Muir and Robert Underwood Johnson, assistant editor of *Century Magazine,* set off on a two-week camping trip in June 1889. The urban-dwelling Johnson had known the mountaineer naturalist for many years. In fact, as editor for *Scribner's,* in 1877 Johnson was so enamored with Muir's writing that he offered to publish "everything you write."[35] Now the two hiked together in the Sierra Nevada. Like many other wilderness novitiates, Johnson fell under Muir's spell as they camped through the "Range of Light." Johnson later recalled how much he enjoyed the trip, especially the fireside talks, as "we lay by the fire and revealed our inmost selves (as one does only by the fireside) until we were overcome by sleep."[36] On one of those fireside evenings they made a pact. If Muir would write two articles on the Yosemite high country for *Century Magazine,* Johnson would lend his considerable support and influence to the idea of establishing a national park, encompassing about two million acres surrounding the Yosemite Valley. Both fulfilled their side of the pact.[37] With the help of Representative William Vandever of Los Angeles and Daniel K. Zumwalt, a land agent for the Southern Pacific Railroad, the bill sailed through Congress in an amazingly short time. President Benjamin Harrison signed the Yosemite National Park bill into law on October 1, 1890, just over a year after the fateful camping trip.[38]

The new national park included the Hetch Hetchy Valley, while, ironically, its sister, Yosemite Valley, languished under state control. The status of the private landholdings within the Hetch Hetchy Valley—often designated as "inholding" by the National Park Service—were not affected. Significant valley homestead lands would give the city of San Francisco an opportunity to purchase ownership in future years. When President Grover Cleveland established Stanislaus National Forest in 1897, to the north and west of Hetch Hetchy, he extended federal jurisdiction, while acknowledging a mosaic of private and public lands.[39]

WHILE THE FEDERAL government worked on land ownership patterns in Yosemite National Park, an important demographic movement swept across the United States that would indirectly, yet significantly, influence the fate of the Hetch Hetchy Valley. The rise of the city brought industry and individual wealth but was accompanied by growing pains. A nation that had traced its roots to the farmhouse was now moving to city tenements. Californians who had gauged their lives by nature's seasons and the rising and setting sun now found life's metronome in the repetitive regularity of the factory whistle and the time clock. The harsh sounds of industry and mechanization replaced the more dulcet sounds of nature. Furthermore, white Americans on the West Coast felt anxious, if not threatened, by immigrant groups such as the Chinese and—soon to come—the Japanese. Not all those who fought for preservation of natural areas were reactionary and racist, but it is fair to say that many were concerned, wondering if all this change was for the better. Their anxiety was enhanced by writers such as Ernest Thompson Seton, Theodore Roosevelt, and Frederick Jackson Turner who lamented the nation's relinquishment of its roots, suggesting that one result might be the deterioration of American character, so strongly influenced by the frontier and so intimately connected with nature. The American world of the nineteenth century was vanishing, and it seemed as if only the landscape artists might retain it for the nation's memory.[40]

The rise of the city was something of a paradox. Commenting on Hetch Hetchy Valley, the historian Peter Schmitt noted that "the urban migration . . . might well have created a water shortage in San Francisco, but it also increased federal responsibility to keep the public domain in trust for those who visited the valley."[41] On one hand, the burgeoning population of San Francisco and the Bay Area towns placed pressure on city officials to find more water. Hetch Hetchy provided one alternative because the valley offered a perfect storage receptacle. On the other hand, many believed that the evils of the growing city could be offset by more parks and places of solitude. People needed nature to counteract the debilitating effects of urban life.

Responses to the rise of the city were complex and not easily defined. Many Americans found Dartmouth professor Liberty Hyde Bailey's "back to the land" idea appealing, while nature writers such as John Burroughs, with his comforting rural ways, enjoyed surprising popularity. However, the people who fought to save the Hetch Hetchy Valley, it must be emphasized, had no desire to turn their backs on the city. On the contrary, the great majority had little interest in Bailey's or Burroughs's admonitions. They lived in the city and were not tempted to leave its comforts and culture. They read the graceful writing of such authors but rarely acted on their earthy philosophy. What motivated them was not the agrarianism of the "back to the land" movement, but rather a desire to go "back to nature" in the sense of a temporary retreat from what Robert Marshall, the New York City resident with a passion for wilderness, would later call "the terrible neural tension of modern existence."[42] They viewed mountain wilderness as a vacation spot, not as a workplace. They were of the middle class, educated, and with no pecuniary interest in the Sierra Nevada mountains. They were little different from Eastern hikers and members of the Appalachian Mountain Club who had "no fuzzy nostalgia for a lost agrarian ideal."[43] Rather they were city dwellers who believed that frequent exposure to nature was one means of enhancing and adorning their urban lives. They viewed Hetch Hetchy as a leisure resource, where they might reflect, perhaps, on the sins of the city before returning to it.

The Sierra Club, founded by Muir in 1892, mirrored this urban bias. Although Muir's core wealth came from tending his wife's family orchards, he never espoused the agrarian life, and he would agree with the assessment of his wife, Louie Strentzel Muir, that "a ranch that needs and takes the sacrifice of a noble life, or work, ought to be flung away beyond all reach and power for harm."[44] The 192 Sierra Club charter members concurred. They included lawyers, professors, scientists, teachers, preachers, and a number of other professional men and women who espoused William Smythe's belief that the good life blended "the cream of the country and the cream of the city."[45] Their "country," of course, was not the pastoral orchards of the Central Valley but rather the Sierra Nevada mountains. In Muir's Range of Light they would find sylvan nature, a place of pleasure and contemplation. They committed to saving the Hetch Hetchy Valley for their recreational enjoyment and spiritual enrichment, with no thought of making money. The defenders were sufficiently affluent that they could afford to view the valley without harboring thoughts of water for irrigated crops, timber for housing, or lamb for dinner tables. Prominent conservationists such as Muir, Robert Underwood Johnson, Harriet Monroe, and countless others believed in "the healing power of Nature."[46] Visitation to a valley such as Hetch Hetchy, a kind of baptism in nature, could be a cure for the maladies of the city, to be taken once a year with or without a community of fellow campers.

The cure for the affliction of the city took many different forms. San Franciscans might soothe their angst through a picnic in Golden Gate Park or a stroll on North Beach. Outdoor recreation, whatever the form, became more popular throughout the United States, and increasingly the "indoor America, sedentary and dyspeptic of disposition, was giving way to a new American."[47] Those who fought for Hetch Hetchy always praised the valley as a scenic place to camp. Camping, once considered only a necessity, had become popular. By the last quarter of the century, middle-class people "experimented with a variety of types of camping—some solitary and physically rigorous, some communal and less demanding." Whatever the case, in the words of historian Cindy Aron, "camping seemed to fit perfectly the needs of a growing vacation public. It promised health, rest, and enjoyment—all for a modest price."[48]

Americans throughout the nation enjoyed more leisure time. New Yorkers retreated to the Adirondacks while San Franciscans found the Sierra Nevada alpine meadows to their liking. This should not suggest that Americans were trying to emulate Joe Knowles, the publicity-conscious "wild man" who allegedly stripped naked to survive the rigors of the Maine woods and thus became a minihero. Nor were they prepared to join Muir, who "might go on foot to the Sierra Nevada, and make his bed in a hollow tree and his dinner of tea and a crust of bread."[49] If there was a "wilderness cult," it was a cult of the mind, not the body. Many avid readers of Muir enjoyed his words and books from the comfort of their home hearths, experiencing the wilderness through Muir's rapturous descriptions and harrowing tales. Few wanted to experience firsthand what Muir could give them vicariously.

Those campers who did visit the Hetch Hetchy Valley did not seek hardship and hunger. Rather, they wished to savor a close but comfortable experience with nature. They differed only in their locale from other hiking clubs across the nation. The Appalachian Mountain Club (1876), the Rocky Mountain Club (1875), the Mazamas of Portland (1894), each promoted the communal and social aspects of camping. They represented a view of nature that did not necessarily force the participant to abandon convenience or culture. The Sierra Club annual outing, beginning in 1901, became a popular event for both women and men, featuring adventure, cultural and moral uplift, romance, and a chance for club leaders to proselytize for nature and the national parks. This new phenomenon of camping and what one historian has called "the golden age of hiking" created a constituency that would be heard in the halls of Congress.[50]

Before San Francisco transformed Hetch Hetchy, camping was the primary use. Like the Central Miwoks' and the sheepherders' use of the valley, it was low impact. Although participants might not acknowledge the fact, communal camping was another form of money-spending tourism. One hundred and fifty people could have an economic impact as well as an environmental one. Each camper in a Sierra Club outing to Hetch Hetchy paid a significant sum of money, much as those involved today with ecotourism packages do. The club encouraged visitation to Hetch Hetchy, and if it might realize a modest profit from the venture, so much the better. Aside from dues and voluntary contributions, the fee was the only way to support the club's political efforts.

Meanwhile, the Yosemite National Park boundaries made little sense. Like many hurried land decisions, Congress in 1890 established park boundaries along longitude and latitude survey lines with little thought to scenic quality, watersheds, mountain ridges, or property ownership. Owners of cattle and of timberland to the west of the Yosemite Park boundary made management impossible. With legal access to any "inholdings," property owners found it absurdly simple to exploit the park. One cattleman owned a 300-acre "inholding" but allowed his livestock to roam over 45,000 acres of park land.[51]

Yosemite's boundaries needed to be realigned. In the 1890s California congressmen introduced bills to that effect, with the support of U.S. Army officers who found the peripheral boundaries a management nightmare. A lumber company forced the issue in 1903, when it began cutting timber on its properties within the park. The following year Interior Secretary Ethan

Allen Hitchcock appointed the Yosemite Park Commission—consisting of Hiram M. Chittenden, a talented army engineer who had served as superintendent of Yellowstone National Park; Robert Marshall of the U.S. Geological Survey, and Frank Bond of the General Land Office—to make recommendations. In the summer of 1904 the Chittenden Commission took to the field, visiting Hetch Hetchy as well as other scenic sites. It took testimony and was particularly anxious to hear from John Muir, who chafed against seeing any land sliced away, particularly in the alpine country to the south. In the end, the committee recommended elimination of much land that contained marginal scenery, as well as park land honeycombed with private holdings, essentially indefensible against invasions from livestock, incursions by miners, or cutting by loggers. Congress essentially adopted the Chittenden Commission report, and the Yosemite Act of February 7, 1905, shrunk the contours of the park by 542 square miles. Clearly, as one historian put it, the act accomplished the purpose of stripping from the park "any lands limited in natural wonders but rich in natural resources."[52]

Jurisdiction in the Hetch Hetchy Valley again went unchanged, but this congressional action trimmed the western park boundary so that there was only a narrow buffer between the valley and Stanislaus National Forest. More significant, the Chittenden Commission added 113 square miles to the northern part of the park. This land was almost all within the Tuolumne River watershed. To justify its action, the commission contended that the addition was largely for protection of water rights. But rights for whom? The report admitted that "already a large portion of its [the Tuolumne River's] water is appropriated and the time may soon come when municipal needs will further draw upon them." Such a statement can be seen as a dark cloud over the valley's future. Aware of San Francisco's desires, the commissioners seemed intent on protecting the city's water interests from private entry. They did not go so far as to recommend the use of Hetch Hetchy Valley as a reservoir site, but they did suggest that national park land could be beneficial to municipal or state water users.[53]

Another ominous sign for Hetch Hetchy was congressional approval in 1901 of the Right-of-Way Act. This legislation opened the park to municipal water exploitation by authorizing rights of way across California parks, as well as other government reservations in the state, for electrical plants and transmission lines, for telephone and telegraph lines, for canals and pipes, and as one critic put it, "for about any purpose whatever." It was "perfectly tailored for looters of the parks," although the secretary of the interior did have the power to review requests and refuse them if justifiable.[54]

To its credit, the Chittenden Commission recommended repeal of the

1901 Right-of-Way Act, but Congress ignored the suggestion. Without doubt the legislators passed the 1901 Right-of-Way Act to aid the water interests of San Francisco, and the idea of repeal was firmly resisted. But still the act's origins are shadowy at best. We do know that Representative Marion De Vries of Stockton, California, pushed the bill through the lower house of Congress, while Senator George C. Perkins shepherded it through the upper house. It went through Congress with virtually no debate or opposition. But who wrote it and put it into the hands of De Vries? That question will never be answered to anyone's satisfaction, but the fact that in 1901 Mayor Phelan had boosted his interest in water and was surreptitiously entering water rights claims in the Hetch Hetchy Valley gives pause. Joseph B. Lippincott, a water engineer of questionable ethics, assisted him. That the Right-of-Way Act passed in the same year as the city moved to claim the valley provides strong circumstantial evidence that the two events were connected and perhaps premeditated.[55]

The passage of this act, an ominous entrée for exploitation of Hetch Hetchy, did not concern the Sierra Club at the time. A tiny organization without any Washington, D.C., connections, the club focused its limited energy toward Sacramento. In the California capital, Muir and the preservationists worked to convince the state legislature to return Yosemite Valley to federal control. There was great satisfaction when they finally accomplished the task in 1906.[56] In the meantime clouds were gathering in the remote valley. Neither Muir nor the Sierra Club, nor the handful of persons who had camped in Hetch Hetchy, suspected that its smooth flowing waters and lush meadows would soon be endangered. For Muir the valley had changed little since his 1871 solo visit. He loved this place for its solitude, its ancient groves of oak and pine trees, its plunging waterfalls, and its riotous meadows deeply set in granite frames. Protected by the umbrella of Yosemite National Park, the valley of glacial origins could only give him satisfaction. This "great landscape garden, one of Nature's rarest and most precious mountain temples," as he described it, seemed secure.[57] In spite of a progression of human use, the valley had retained its integrity. It seemed safeguarded under Muir's watchful eye and the protection afforded by the U.S. Army. However, no one had reckoned with the ambitious leaders of San Francisco, who had no acquaintance with or feel for Muir's valley or, if they did, considered its loss only a small price to pay for urban progress.

CHAPTER 2

The Imperial City and Water

"Cities, like all other living things, need water to survive, and even more water to flourish."

GERALD T. KOEPPEL, *Water for Gotham: A History* (2000)

IN THE SAME year that Joseph Screech wandered into the Hetch Hetchy Valley, San Francisco was in the midst of a remarkable change from a city of shacks and tents to one of wood frame houses and commercial enterprises. In the brief time from the American takeover to the gold rush, a mere three years, the Mexican town of Yerba Buena transformed itself into a commercial center of some 40,000 people, a cosmopolitan mixture of ethnic and racial groups, all driven by a desire to acquire wealth in this new country—and all in need of water.

From this beginning, water in San Francisco was of paramount importance. In many regions residents assume abundant water as a given right, but it has always been problematic in the arid West—highly valued, often scarce or unavailable when you want it. San Francisco was on the edge of aridity, but its location made it particularly vulnerable. So the city sought to solve its resource problem by looking to the snow-fed streams of the distant Sierra Nevada mountains. City leaders understood that availability of water often influenced urban growth as well as individual fortunes. For human communities water is a necessary commodity that has multiple meanings, often bringing out the best and most noble human characteristics but also quite capable of triggering self-serving actions and greed. In the water wars of San Francisco, all of these traits were present.

The gold rush culture that emerged by 1850, dominated by Americans and built on a "pyramid of mining," overwhelmed the indigenous peoples as well as the Spanish and Mexican residents. Yet the importance of a reliable supply of water did not originate with the American takeover of San Francisco. The Indians of California had long manipulated their environment, including diverting their source of water. However, in most cases, the Indian peoples, such as those in the Hetch Hetchy Valley, chose to live near water sources, rather than divert or store them.[1]

Spanish settlement in Alta California in 1769 brought a new awareness of the necessity of water. The Franciscan friars and the presidio soldiers were familiar with arid lands and, as they established missions along the coast, took great care to plan and build with water in mind. They based their economy not only on grazing but also on intensive agriculture. And if they succeeded in converting the Indians to Catholicism, a growing population would demand a bountiful supply of the precious commodity. Furthermore, the Spaniards believed that control of water could be viewed as a foreign policy tool, enhancing their settlements while discouraging other European powers with designs on this faraway region. Water, as historian Norris Hundley has noted, was "crucial to Spain's successful colonization of California . . . as a means of imperial expansion."[2]

In spite of the care they gave to water sources, the padres made mistakes in site selection, resulting in either flood or drought. The Mission San Carlos Borromeo, Father Junípero Serra's headquarters, had to be relocated for this reason. Usually "prayer rather than relocation became a frequent response to capricious weather cycles," but occasionally the Franciscan padres became embroiled in controversies that appeals to the Almighty could not resolve. The Mission San Fernando and the pueblo of Los Angeles failed to agree on an equitable management of Los Angeles River water, necessitating negotiation and compromise. In Northern California the rather insignificant Guadalupe River became a source of contention between the Mission Santa Clara and the pueblo of San Jose. Since populations were small, normally the disputants settled their differences amicably, although the later controversy lasted for two decades and caused bitterness between the Hispanic and Indian communities, as well as the padres, who generally sided with the Indians.[3]

Although the mission's padres defended Indian water rights, one should not presume that the first Californians benefited. Water projects, then as now, required labor, and the mission Indians supplied it. The padres, of course, never considered the neophytes as slaves, yet they did force them to construct the primitive hydraulic works necessary for expansion of irrigated land. The building of an irrigation system served a dual purpose: It provided a needed crop infrastructure while controlling the Indians and assimilating them to intensive agriculture.[4] Such structured labor, so culturally foreign to the natives, caused alienation at best; at worst, death.

In the case of water disputes, Spanish-Mexican authorities almost always privileged community needs over individual rights. Water was an invaluable resource to be shared, held in common for the good of the community. However, with the American takeover the political culture changed dramatically. American newcomers were confident, impatient, and arrogant, and armed with their capitalist ideas, they felt unconstrained by limitations.[5] They believed that the abundant natural resources of California, whether grass, timber, precious metals, or water, were at their disposal—all in the name of progress, or more specifically, empire building on the Pacific Rim. They would extend no special community status to water. If you were clever enough, it could be privatized for profit. Men such as William Ralston, William Sharon, and George Hearst were hardheaded businessmen, often greedy, uneducated, and embracing a cut-throat code that was the only one they understood.[6] In a sense, they represented the failure of idealists such as Emerson or Thoreau to produce any sort of social philosophy that embraced community or encouraged respect for the limits of nature.

The exploitation of California's natural resources continued unabated in the years leading to Hetch Hetchy. Those who presumed to speak for wealth, much of which flowed to San Francisco, believed they were transforming a pioneer land into a settled, civilized one. And in a large sense, the waters of California served as the converting agents. Logging companies, which cut the coastal redwoods as well as the sugar pine forests of the Sierra Nevada, often used the energy of river waters to transport logs to mills. The most common work of water might be found in the dominant mining industry. From simple pan placer mining, to river mining, to quartz mining, to hydraulic mining, all shared in the need for water. However, that need graduated exponentially with hydraulic mining to the point that mining and water bonded and could not be separated. In time, a large, pressurized supply of water became even more important than the gold-bearing gravel and soil that it reduced.[7]

Hydraulic mining provides an example of man's most environmentally destructive activities, one in which greed transgressed all thought of community concern. This form of mining used whole rivers to extract a few ounces of gold for a commercial company. Workers transported water by ditch, flume, or pressurized pipe, which in turn fed into a nozzle so powerful that it made today's fire hose seem like a child's toy. The more primitive nozzles required six or eight adults to hold them. The more sophisticated ones had to be anchored by steel and iron. Hydraulickers washed away mountains of rock, gravel, and debris for gold. They represented the antithesis of the individual placer miner with his simple pan or sluice box, who by 1855 remained only a romantic memory. In his place emerged San Francisco mining corporations, staffed by bank clerks and engineers more familiar with the workings of stock trading than a sluice box. Mining operations required prodigious amounts of capital and water, and the financiers of Montgomery Street in San Francisco usually supplied both, selling water by the "miner's inch," the amount of water that flowed through a square-inch hole in ten hours under constant pressure.[8] Some water companies merged and consolidated, to become so large that they bought out mining companies. In 1856 the Eureka Canal Company in El Dorado County boasted dams, reservoirs, and 247 miles of canals valued at $700,000. By 1865 the restructured Eureka Lake and Yuba Canal Company owned its own mining operations, delivering 85 million gallons every 24 hours to its extended sites.[9] In an age in which electric power did not yet exist and steam power was expensive, the kinetic energy of water dominated.

The technical problems that emerged from hydraulic mining challenged engineers to advance their knowledge. William Ralston's North Bloomfield Mining Company faced what appeared to be insurmountable runoff problems at the Malakoff Mine. Undaunted, hydraulic engineers designed a tunnel some 130 feet beneath the surface, requiring workers to drill and blast a tunnel a mile and a half long through bedrock at a cost of over a half million dollars. To the amazement of skeptics, the design was successful. North Bloomfield was the largest operation, but more than 90 other hydraulic operations sifted gold from a massive amount of soil.

The environmental calamity resulting from these companies hell-bent on profit and expansion staggers the imagination. Mountains disappeared as the companies swung their great nozzles toward any section of earth that had potential to yield profit. Water became the instrument of a business dominated by human greed and unregulated capitalism gone berserk.[10]

In addition to the immense environmental damage, there were social costs as well. During the hydraulic mining era (1852–1884), debris-filled rivers flowed onto the Sacramento Valley. Once there, the water slowed and the silt and debris settled. By 1874 the Yuba River flowed in places sixty feet higher then it had in pre–gold rush days. During spring runoff, desperate farmers and townspeople could not raise levees fast enough. High water broke levees, coating hundreds of farms with debris that residents called "slickens," ruining the land and driving farmers into poverty. One need not have a vivid imagination to understand the problem. Powerful San Francisco mining corporations unleashed vast amounts of debris-laden water into the mountain rivers with no concern for downstream users. Water has and will cause destruction through floods, but such events represent uncontrollable chance acts of nature. This catastrophe represented the deliberate rapacity of men of wealth. San Francisco corporate heads should have said "This is wrong" and stopped or at least mitigated the practice. However, their reaction to the suffering of local communities was to hire attorneys to defend their right.

This mining practice would not have ended except that farmers and valley merchants banded together to form the Anti-Debris Society. They sued for relief but found none in the California courts. Then Edward Woodruff, an absentee landowner, brought suit in 1882 in the U.S. Ninth Circuit's district court in San Francisco, where Judge Lorenzo Sawyer presided.[11] On January 7, 1884, hydraulic mining came to an end. Judge Sawyer, after reviewing more than 20,000 pages of testimony and making numerous field trips to observe the mines and the downstream damage they caused, gave his 225-page decision in *Woodruff v. North Bloomfield* (which he insisted on reading aloud to an impatient audience), calling for a permanent injunction against hydraulic mining.[12] Common sense won out over the industrialists' greed as farmers and the environment were left to recover lives and land. Close to 20,000 miners packed up their gripsacks, and many hydraulic engineers began to look for work. They had the talent to move massive amounts of water. Why not capture the waters of the Sierra Nevada for a beneficial use, rather than glean a few ounces of gold? The engineers could look on the Hetch Hetchy Valley as simply another challenge, one more satisfying than turning mountain waters into chocolate-colored rivers of mud and debris.[13]

The hydraulic mining wars underscored the power of San Francisco, the financial center of the West. Like the water currents of the Golden Gate, commerce surged both east to the mines and west to far-flung seaports, spawned largely by the needs of the mines. Above all, it was the bankers, venture capitalists, and business interests who determined the politics and the economy of the state. According to one writer, this domination of San Francisco took place not by mere chance but by design. Author Gray Brechin likens the rise of San Francisco to the Italian city-states of the fifteenth and sixteenth centuries. Such cities as Milan or Venice dominated the surrounding countryside, or *contado,* which was subservient to their needs. The city could offer the countryside protection from neighboring enemies as well as markets for farm products, while the *contado*'s peasant population provided food, natural resources, labor, conscripts, and even taxes for the dominant city-state. Of course oligarchies, which had no semblance of democratic virtues, controlled the Italian city-states. Yet Brechin sees little difference between such oligarchies and what he terms the "linked dynastic elites" of San Francisco. These wealthy families—such as the Spreckels, de Youngs, Phelans, Bourns, Crockers, Stanfords, and Hearsts—all controlled the flow of information, "editing out that which negatively concerned themselves." These families represented an urban aristocracy dedicated to wealth, power, and above all, growth. They all agreed, says Brechin, "that the city must grow—and its land values rise—to assure the continuation of their dominion."[14] To harvest such growth, San Francisco reaped the resources of the countryside without much sowing. Acquiring gold from the mountains at the impoverishment of farm land below was a perfect example. When San Francisco leaders laid siege to the Hetch Hetchy Valley, they believed it was their right. The Italian city-states were not shy in claiming the countryside, and neither was San Francisco. The city had no standing army, but it did have population, political power, and a certain sense of entitlement.

Still, there were limitations on San Francisco power that no Italian city-state had to endure. Above all, the city had to contend with the federal government. When the farmers got no satisfaction on hydraulic mining from the California legislature or state courts, they went to the U.S. Ninth's district court. There the "hinterlands" received a modicum of belated justice. Later, federal power would force San Francisco to compromise on its desires for absolute control of the Hetch Hetchy water system. The San Francisco plutocrats also had to contend with the fifth estate. A free, uncontrolled press and a vigorous journalistic tradition often led to revealing investigative reporting. Newspaper publishers, such as the de Young brothers and William Randolph Hearst, could be absolutely slanderous at times, and yet their investigative reporters kept politicians and the "dynastic elites" nervous, and to a degree, aboveboard.

It would be a mistake, however, to suggest that only the dynastic elites fought for Hetch Hetchy to feather their nests. The middle class supported the effort, knowing that an improved urban infrastructure would not only enhance their lives but also raise the value of their more modest real estate holdings. It did not matter if their stake was a quarter of an acre, rather than thousands; they understood the principle of growth and appreciating land values. When it came to the Hetch Hetchy water system, people of the Bay Area, regardless of middle- or working-class status, saw the investment in abundant, pure water to be in their interests, although at times they were reluctant to pay for it.

While I have suggested restrictions on Brechin's *contado* idea, one must also grant it credence. San Francisco leaders had a vision, and it included use of the surrounding lands. Through patience, determination, arrogance, and a special sense of mission, they pursued the vision of a great city. San Francisco's determination was similar to that of Chicago, where the word *empire* was on the lips of every city booster, even though such a metaphor was highly undemocratic. Chicago desired hinterland hegemony, and so did San Francisco.[15]

From the viewpoint of James Phelan, a man steeped in the classics, such cities as Babylon, Thebes, Athens, Alexandria, Carthage, and Constantinople achieved greatness through their infrastructures and control of the surrounding countryside. Rome, of course, was never far from his mind. He admired the architecture and transportation system, but it was the aqueducts, furnishing the city with abundant freshwater from the hinterlands, that offered the most pertinent example for San Francisco. The greatness of Rome depended on its aqueducts. Phelan admired their engineering, but he knew that construction had not come quickly. For four and a half centuries the city depended on local wells and the Tiber River. But the empire needed more water, and between 312 B.C. and A.D. 226 engineers designed—and slaves built—eleven gravity-fed masonry aqueducts, bringing some 38 million gallons of freshwater a day from distant lakes, rivers, and streams. In its glory days, the days Phelan and San Francisco boosters liked to recall, ancient Rome's abundant water supply helped create an almost aquatic culture that depended on the public and private baths, where Romans gathered for cleanliness, enjoyment, and discussions of politics and business.[16] No doubt Phelan committed to memory the story of how the Gothic leader Vitiges and his army blocked the aqueducts leading to the city, thus ending the imperial rule of Rome forever.[17] Clearly civilization, and indeed survival, depended on a bountiful water supply. Not surprisingly, when architects designed the two elegant water towers associated with the Hetch Hetchy system (Sunol and Pulgas), one replicated the water tower of Tivoli that fed Rome, while the other followed a Roman Renaissance revival architecture, modern proof that symbolism plays an important role in the perception of a city and its countryside.

THE FIRST SYSTEMS to provide potable water to the city of San Francisco bore no resemblance to the magnificent aqueducts of Rome. Residents in 1848 acquired water by the most primitive means. With no viable streams close at hand, owners of homes, restaurants, hotels, and shipping offices dug crude wells, often brackish or dry. A more reliable source was the water boats that plied daily across the Golden Gate from Sausalito and Marin

County, where streams coming off Mount Tamalpais offered a sufficient supply of drinkable water. Water venders unloaded the cargo and then negotiated the dirt (or mud) streets with their carts, selling buckets and casks of what might truly be called "liquid gold." For a time the carts quenched the thirst of San Franciscans, but no wonder the numerous bars were rumored to compete with the water carts for sales.

This primitive water system gave no protection against fire. The ramshackle town of San Francisco, whose population far exceeded its infrastructure's capacity to provide for it, burned down *seven* times between late 1849 and 1851.[18] Many criticized the inadequate volunteer fire department and the lack of water and equipment. However, the limited water supply was unfairly blamed. Often fires were the work of unruly gangs, such as Australian immigrants known as the "Sydney Ducks" who were not above using arson to create havoc, allowing them to pillage property.[19]

As the gold fever subsided, San Francisco looked to replace the water carts with a respectable infrastructure. In 1851 the Mountain Lake Water Company contracted with the city to bring water by flume from Mountain Lake and Lobos Creek, both on the sand dune–dominated western edge of the city. The company never succeeded and, plagued by lawsuits, went broke in 1862. In the meantime entrepreneurs formed the San Francisco Water Works in 1857. This company also tapped Lobos Creek and succeeded in bringing water by flume around the shore of the Golden Gate and into the city, where pumps elevated it to a suitable level for distribution. A year later, George H. Ensign organized the Spring Valley Water Works, developing a small spring in the city, close to the intersection of Mason and Washington streets. He laid a few pipes but, lacking capital, he was bought out by a stronger consortium in 1860 that retained the *Spring Valley* name. In 1865 the San Francisco Water Works and the Spring Valley Water Works consolidated. The Spring Valley Water Works, which changed its name to "Spring Valley Water Company" in 1904, became the dominant water utility in the city. No other company in California was so controversial, save perhaps the Southern Pacific Railroad, and no other private enterprise was so involved with the fate of the Hetch Hetchy Valley.[20]

Although the new company furnished the city with an adequate supply of water, it soon embroiled itself in controversy. Most of its disputes, often in the courts and always in the San Francisco newspapers, centered on public ownership and water rates. Hardly had the company formed when city supervisors appointed a special committee to eliminate it. The committee of General B. S. Alexander of the Army Corps of Engineers and Professor George Davidson of the U.S. Coastal Survey were to evaluate the water supply needs of the city with an eye toward municipal ownership. In 1871 they recommended that the city ought to have "absolute control" of its water supply. The committee noted that in other cities such as New York, Boston, Philadelphia, Chicago, and Washington (and soon Los Angeles), the water system was under municipal ownership. San Francisco should do no less. In 1877 the supervisors took action, offering to purchase Spring Valley for $11 million. The company demanded $16 million, and the gap could not be bridged. Had the two sides arbitrated the difference, many years of controversy and charges of cunning and corruption could have been avoided. Furthermore, the fate of Hetch Hetchy Valley might have been very different. More than 50 years ensued before the city of San Francisco finally passed the bonds to buy out the Spring Valley Water Company.

Part of the difficulty in purchasing the company involved the ownership. From the beginning the stockholders were men of wealth, part of the "dynastic elites." San Francisco financiers such as Charles Crocker, Leland Stanford, Mark Hopkins, Collis Huntington, Darius Ogden Mills, William Ralston, James Ben Ali Haggin, and Lloyd Tevis all held interests in the Spring Valley Water Works.[21] Even the famed landscape architect Frederick Law Olmsted recognized this valuable stock, although he had nothing but contempt for most of the owners. When he arrived for his two-year stay in California (1863–1865), he soon purchased Spring Valley stock.[22] Why not? The utility company kept acquiring both land and water rights, and increased its return on investment. In 1870 the company grossed $817,000 from its customers, returning $480,000 in dividends, a 59 percent return. Certainly a smile must have crossed the Spring Valley nabobs' faces when, in 1875, the company returned a 61 percent yield.[23] Wealth begets wealth, and though mining fortunes would shrink later in the century, unregulated private utilities offered lucrative opportunities.

Alongside utilities, the main opportunity for the wealthy shifted from the roller-coaster world of mining stock to more conservative real estate investments. By 1885 William Randolph Hearst recommended to his father, George, who had amassed his fortune through mining ventures, that he direct his investments to land. With second-generation wisdom, William declared that "the landlords are always a wealthy class." Each child born created another hungry mouth to feed and "every atom of humanity added to the struggling mass means another figure to his [the landlord's] bank account."[24] His jaundiced analysis of society aside, the owner of the *San Francisco Examiner* realized that population growth demands space and that land would ultimately appreciate.

This fact was not lost on the stockholders of Spring Valley either. Certainly water could yield amazing profits, but land could yield even more *if* the owner could ensure water. It was, and is, the marriage of land and water that gives land its value, particularly in the American West. No one knew this better than Harry Chandler, James Otis, Fred Eaton, and William Mulholland, the men who worked and schemed to bring Owens River water by aqueduct to the parched land of the San Gabriel Valley, thus laying the foundation for the growth of Los Angeles, and also lining their pockets. Of course the San Francisco Peninsula was better watered than the San Gabriel Valley; yet 20 inches of rain, when it all came in a winter deluge, was not sufficient. Coastal fog slipping over the hills could not relieve the dry summer months as the creeks vanished and grasses turned brown, or as California boosters preferred, "golden."

A dependable, regulated water supply for not only San Francisco but the southern lands as well could be profitable in the long run. This reality was evident to Spring Valley's talented Swiss engineer, Hermann Schussler. Under his leadership in the 1860s, the company aggressively expanded into the watersheds of San Mateo County. First came the purchase of the Pilarcitos Creek watershed, diverting the creek's flow from the Pacific Ocean to a redwood flume aimed at San Francisco. More purchases followed, particularly through the heart of the scenic San Andreas Valley, which stretched north to south for some 25 miles. Once the company agents secured the land, Schussler laid out the reservoirs of San Andreas and Crystal Springs. By purchase or eminent domain, the company increased its holdings to more than 100,000 acres, building a water empire as it excavated earth, siphoned water, and laid pipe.

Schussler possessed exceptional engineering ability. In 1873 he supervised the laying of an aqueduct that stretched from Lake Tahoe across the level lands of the Washoe Basin and then rose to Nevada's Virginia City. Many thought this gravity system would be a disaster, but it worked. Schussler was one of those hydraulic engineers quite determined to "reconfigure the planet's plumbing."[25] He enjoyed the challenge of moving water to where it was wanted and relished shaping his environment to meet his engineering objectives. No job seemed impossible for his hydraulic engineering skills. His capstone was the construction of the Crystal Springs Dam, completed in 1890 and reported to be the world's largest concrete dam at the time.[26] The structure captured the waters of San Mateo Creek, creating Crystal Springs Reservoir. Ironically, Schussler did not know that directly beneath his reservoir the Continental and Pacific Rim tectonic plates ground against each other, creating a frightening tension and energy that would be unleashed on the city of San Francisco in April of 1906. However, perhaps as a tribute to his engineering skills, the great earthquake did no structural damage to his 150-foot-high dam of interlocking concrete blocks. A century later it remains solidly in place.

Building the Schussler system for Spring Valley was not without controversy. Company agents purchased land secretly and often under false pretenses. Reluctant landowners found themselves in court, facing company attorneys who claimed the right of eminent domain. The owners might also have to defend themselves against the San Mateo County supervisors, who were often beholden to Spring Valley. In the southern part of the county angry landowners often received only a fraction of their property's worth. The company became a bully but justified its aggressive style by citing San Francisco's insatiable water demands. But that was not always the case. When the residents of the tiny town of Searsville, to the west of Palo Alto, lost their court case in 1879, they saw their homes inundated by fathoms of water. Ironically, the California State Railroad Commission determined in 1892 that Spring Valley could do without the Searsville reservoir water. Stanford University ultimately acquired the 105-acre lake for irrigation, recreation, and research purposes.[27]

San Francisco watched the expansion of the Spring Valley system with ambivalence. On one hand, the city needed more and more water, and thus growth drove expansion. On the other, city officials dreaded the company's political power, particularly in setting rates. The city seemed to be held hostage by Spring Valley. This was particularly true in the 1870s when the company accrued excessive profits through its monopoly and its exploitation of a laissez-faire system. However, that situation changed when the California constitution of 1879 empowered the San Francisco Board of Supervisors with the responsibility of annually setting water rates.[28]

Certainly the supervisors curtailed Spring Valley's exorbitant profits, but the new constitution could do nothing to curb city corruption. San Francisco had been no stranger to urban graft, but it seemed to become institutionalized by the 1880s. As in New York and a number of other cities during the gilded age, city government fell into the hands of political bosses who wielded great power without the moral or ethical leadership one might hope for in a democratic system. San Francisco had its own handful of bosses to rival the notorious Boss Tweed of New York's Tammany Hall. In the 1880s "blind boss" Chris Buckley held sway over both the Democratic Party and City Hall. Republican Martin Kelly replaced him. Mayor James Phelan gave the city a splash of honest government, but by 1902 Abe Ruef, a refined former valedictorian of the University of California who evidently had taken no courses in ethics, headed the world of rake-offs at City Hall. Only after a judge sentenced Ruef to 14 years in San Quentin prison, in a political shake-up almost as violent as the 1906 earthquake, did some semblance of honesty return.[29]

The modus operandi of the bosses was to receive "retainers" from Spring Valley as well as other municipal monopolies, and then to distribute bribes to supervisors for necessary votes. The pattern began immediately after the supervisors were entrusted with setting water rates. In 1883 the San Francisco electorate voted in Washington Bartlett as mayor on a platform committed to lowering Spring Valley rates. Supervisor J. J. Reichenback supported him, declaring that "justice demands this." However, Mayor Bartlett did not take into account "blind boss" Buckley's annual Spring Valley "retainer" of $5,000 to $10,000. When the supervisors passed the 1883 water ordinance with no reduction in rates, Bartlett discovered how powerless he was. One week later the supervisors passed the ordinance over the mayor's veto, in a vote of nine to two. Among the majority was Supervisor Reichenback.[30]

By the mid-1890s the city of San Francisco cleaned house with the election of Mayor Phelan. Also, Spring Valley reversed its corrupt practices under the direction of William Bourn. Both men were of high principles, and each led his organization in more respectable directions. Yet there was little Bourn could do to stem the tide of Spring Valley's bad publicity, which seemed to continue unabated from 1880 until its sale to the city in 1928. It was a private monopoly at a time when almost all American cities' water supplies were municipally controlled. In fact Spring Valley was the largest investor-owned water utility in the nation. With its large landholdings in five counties, representing an area three times the size of San Francisco, Spring Valley seemed an "octopus" only exceeded in its reach by the equally despised Southern Pacific Railroad. The company seemingly proved the axiom that you cannot outlive your past. When the San Francisco earthquake turned into a fire that devastated four square miles of downtown, it was Spring Valley that absorbed much of the blame. One San Francisco magazine, *The Grizzly Bear,* labeled Spring Valley an "arrogant and consciousless monopoly" that had cost the city $500 million.[31]

The Grizzly Bear, as the magazine's title suggested, grew out of San Francisco's scrappy, devil-may-care journalistic tradition. Newspapers, which emerged and disappeared with frightening speed, seemed to thrive on character distortion and rumor. Spring Valley always seemed to offer fodder for such papers as the *San Francisco Chronicle,* which took pride in its reputation as a courageous, vindictive, spunky, master-of-insult daily. The *Chronicle* in many ways assumed the character of its owners, Charles and Michael de Young. For an idea of how inflammatory the paper could be, one must retreat to 1879, when Baptist minister Izaac Kalloch ran for mayor under the banner of the Workingmen's Party. The de Young brothers attacked his candidacy in the pages of the *Chronicle*. No shrinking violet himself, Kalloch, in a public address, called the de Young boys "the bastard progeny of a whore born in the slums and nursed in the lap of prostitution." This description did not sit well with Charles, who took to the streets and from a horse-drawn cab fired his pistol at close range on Kalloch. Injured, the minister survived to fight on and become mayor. Charles did not fare so well. Kalloch's enraged son sought him out and killed him with a well-placed bullet to the throat. Perhaps indicating the city's mood toward Charles de Young, a jury hastily acquitted young Kalloch on the "ground of reasonable cause."[32] Thirty-year-old Michael took over from his deceased brother, running the *Chronicle* for the next 45 years.

In the years to follow, the Spring Valley Water Company was often the target of Michael de Young's bombastic journalism. He seemed to have learned little from his brother's violent death. Conflict between de Young and Spring Valley came to a head in 1909. That year de Young decided he could control the water company through defiance. To test his assumption, he refused to pay the water bill for his *Chronicle* building. Spring Valley president William Bourn, a man who detested what he considered vile behavior, sent his representative A. S. Crawford—who had actually worked for the newspaper for 17 years—to reason with de Young. The discussion got nowhere, the fiery publisher declaring that the water rates were "nothing short of robbery." On October 8, 1909, Bourn dispatched Crawford for the last time with instructions that de Young be informed that if he did not pay his water bill, Spring Valley would be forced to cut his building off. Michael de Young vowed that if that happened, he would assign a reporter to do nothing but dig up dirt on the water works and "rip your company up the back and wide open." When the time for payment came and went, Bourn ordered the water main to the *Chronicle* building closed. It took only a few hours for de Young to reconsider and pay his bill, but true to his word, the publisher continued his attack against Spring Valley and William Bourn.[33] Time and again the media attacked the company, and of course the *Chronicle* became an ardent supporter of the city's plans for the Hetch Hetchy Valley.

From the tortured relationship between the Spring Valley Water Company and the city of San Francisco, it becomes apparent why the far-off valley called Hetch Hetchy became an attraction, indeed almost an obsession, for city leaders. Over the years the company had committed the sin of gouging and feeding from the public trough whenever possible. Between 1860 and 1880 Spring Valley took every advantage allowed by unregulated capitalism.

The San Francisco Peninsula owners, of which there were many, not only garnered remarkable dividends on their stock but enhanced the value of their landholdings by the availability of water. Soon some would be carving sections of land from their manorial estates to create well-watered and profitable subdivisions. Perhaps representative was Francis Newlands, son-in-law of California senator William Sharon, who developed peninsula land with the Burlingame Country Club as the centerpiece. The parcel was first purchased by William Ralston, then became part of Sharon's holdings. But it was Newlands who would successfully develop expensive, well-watered lots that would split off as Hillsborough, still the residential retreat of the rich. Mining money begat real estate wealth, and certainly Francis Newlands, the future Nevada senator and sponsor of the Reclamation Act of 1902, knew a thing or two about the profitable manipulation of water.[34]

As San Francisco's wooing of a water source continued, the Sierra Nevada looked more and more inviting. In truth, long before Mayor Phelan took office, San Francisco leadership had toyed with the idea of a municipal water supply from the distant mountains. However, the mayors and supervisors seemed bent on entertaining proposals rather than initiating them. Perhaps the first and the most outlandish proposal came from Colonel Alexis von Schmidt, a German military man who had participated in the gold rush and then proclaimed himself an engineer. He had confidence and a disposition that bordered on arrogance, and in some ways his ethics paralleled those of boss Abe Ruef. After one hydraulic failure to transport water to Virginia City, he found sporadic employment with the early Spring Valley Water Works. In 1865 he found reason to abandon his employer to promote a company whose name described its purpose: "Lake Tahoe and San Francisco Water Works Company."[35] His intent, of course, was to drain that jewel lake of the Sierra Nevada to provide water for the needs of the growing city. Never mind that to do so would necessitate a water basin transfer and a lengthy tunnel, as well as despoiling a much admired lake. Von Schmidt was not a great engineer, but his professional skills outweighed his political acumen. His scheme would have reduced the Truckee River to a trickle—a stream as central to Nevada as the Sacramento River was to California. The Virginia City *Territorial Enterprise* suggested that if von Schmidt dared to take Lake Tahoe water, he should bring an "escort of twenty regiments of militia." The newspaper conceded that San Francisco mining moguls "may take the gold and silver from our hill and bind us in vassalage to the caprices of their stock boards" but the waters of Lake Tahoe would not flow west.[36] Von Schmidt assured the Nevadans that they were being unreasonable; there was plenty of water for everyone. However, the vision of a dried-up Truckee River and a bathtub-ringed Lake Tahoe reservoir worked against him. Still, von Schmidt did not accept defeat. In 1871 he convinced the San Francisco supervisors to approve his proposal, but the mayor vetoed it, fearing legal challenges. For the moment, the Lake Tahoe proposal was dead.

In the 1870s San Francisco supervisors continued their attempt to bridge the $5 million gap with Spring Valley. But with negotiations at a standstill, the supervisors, as well as the newspapers, were full of ideas for a city-owned water system. Where to find the water? Most every politician, engineer, and journalist looked to the Sierra Nevada. The *San Francisco Examiner* suggested that with an expenditure of $10 million, the city could construct "a line of pipe of one hundred and fifty miles in length" bringing "the purest water known." Just which source should provide the water was unclear. The South Fork of the American River, Clear Lake (north of the city), the Tuolumne River, the King's River, the Rubicon River (a branch of the American), and Blue Lakes (in the Sierra Nevada mountains) were all candidates. The Spring Valley water system seemed to draw less attention, for San Francisco had set its sights on the distant Sierra Nevada rivers. In fact the *Examiner* declared that "the water of the mountains is to that of the Spring Valley Company as is a ripe, sound peach to a rotten one."[37] The *Examiner* advised that the city disregard the cost. London failed to go to its hills and "prefers to pump what it needs from the filthy Thames." San Francisco "can do better."[38]

Across the bay the Oakland newspapers expressed interest in a regional water system with a Sierra Nevada source. The *Oakland Transcript* suggested "that by going into a partnership arrangement with San Francisco, this city, Sacramento, Stockton and other places could get water from the mountain lakes and streams at a very modest cost." One year later the *Oakland Daily News* reported a strong possibility that the San Francisco supervisors "will purchase the water rights of the Mount Gregory Water and Mining Company, and bring the pure waters of the mountain lakes of the Sierras down through Oakland and pass them under the bay." If so, Oakland would have "an ample supply of pure water, and at low rates."[39] Even at this early date, a regional water system was a possibility, and the Hetch Hetchy Valley was not the only candidate. Also evident was that the Bay Area communities looked to San Francisco for leadership, perhaps even dependency, as a child to a parent. This was not surprising, given how the city dominated in population and wealth.

Perhaps because of its confidence in dealing with Nevada's protests, San Francisco resurrected von Schmidt's idea in the 1890s. But now there was a new consideration to the Lake Tahoe scheme, almost as important as the water itself. In the late 1880s the advantages of electricity became evident, and within a decade nascent power companies formed throughout the nation. California was noticeably devoid of coal, and therefore the falling water of the Sierra Nevada became the electrical generating source of choice. "California White Coal" sped the state to the forefront of hydroelectric technology, as well as long-distance electrical transmission from the mountains to the more populated Bay Area and coastal towns.[40] This new aspect of water transfer gave the Lake Tahoe idea new appeal. Tahoe water, once tunneled to the western slope, could be manipulated in a drop of some 7,000 feet to produce thousands of kilowatt-hours of electricity per day, before being transported by aqueduct for use by the city.

In 1900 the city sent an inspection team. The plan's possibility was enticing, particularly since von Schmidt, now desperate, was prepared to sell his water and property rights for $50,000.[41] However, nothing had really changed with the state of Nevada in the intervening 20 years. If approved, the plan would be tied up in the courts, filling only the bank accounts of water lawyers rather than the reservoirs of the city. The inspection team recommended that the city continue to look for sources on the western slope of the Sierra Nevada.

As the city continued its search, the Tuolumne River seemed to surface in the conversations as the most desirable. Although reservoirs might be built downstream, the isolated Hetch Hetchy Valley, at an elevation of 3,800 feet, had the advantage of providing a reserve for hydropower production as well as a storage basin for pure Sierra Nevada snow water. Furthermore, the U-shaped glacial valley had just been made part of Yosemite National Park in 1890. For San Francisco this was a significant event. National park status closed off private land entry, and the few private properties within the valley could be purchased or, if necessary, acquired through eminent domain proceedings. Other viable sites had been rejected by the city engineers and attorneys because they offered a briar patch of legal entanglements and expenses. Hetch Hetchy might require difficult, indeed original, civil engineering, but such physical obstacles seemed hurdles easier to jump than the multiple private claims on land and water that attended other rivers and sites. It would prove a questionable assumption.

Thus it was that by the late 1890s San Francisco had made two significant decisions. The city should have a municipally owned water system, and water should come from a source in the Sierra Nevada mountains, preferably the Hetch Hetchy Valley. All that the city needed was enabling legislation, the acquisition of land and water claims, and a determined mayor. In 1900 they would all come together.

CHAPTER 3

Water, Earthquake, and Fire

"I saw the hillside covered by homeless people. I saw such suffering as I never expect to see again, and I know that a lot of it was caused by reason of the inefficient system of water that was being supplied to the people of San Francisco."

SENATOR KEY PITTMAN

WHEN MAYOR-ELECT James Phelan took office in 1896, he immediately drew wide publicity by taking the law into his own hands. With the blessings of a judge, he threw out eight elected city supervisors, known as the "Boodling Board of Supervisors." Only four remained, whom he presumed to be honest. The disgraced supervisors did not leave without protest. They entered City Hall on a Saturday night, determined to hold their ground until the Monday night supervisor's meeting. Not to be outdone, on Monday morning Phelan arrived with the police. As one reporter noted, the officers forcibly removed the trespassing supervisors, "bag and baggage, like a porter kicking out into the streets a lot of old ash cans."[1] The supervisors of questionable character eventually went to court and regained their seats, but it did not matter. Phelan had made a statement. He intended to be strong and honest, taking government to incorruptible heights in a rarified atmosphere the city had never known. And, of course, he determined to provide San Francisco with a new water supply, pure and owned by the city.

Across the Bay, John Muir's involvement in politics might not have been so dramatic, but he was nevertheless effective. The mountaineer and man of national reputation wished to save his Yosemite Valley from the clutches of a state commission that he felt was every bit as incompetent as Phelan's city supervisors. Yosemite Valley, Muir believed, should be returned to the jurisdiction of the federal government. After lobbying President Theodore Roosevelt on a four-day camping trip in 1903, and with continual arm-twisting of Sacramento legislators, he achieved his goal.[2] At the turn of the century Phelan and Muir were on separate tracks, each pursuing his dream: one for honest government, the other for Yosemite Valley.

If one had to characterize the Hetch Hetchy Valley fight through two personalities, Phelan and Muir would be reasonable choices. Phelan, determined to have the valley, was rich, powerful, honest, cultured, committed, with a sense of noblesse oblige for his city. His father had made a small fortune in the liquor business, a booming enterprise during the gold rush days, and had branched out into trade and banking. James, born in 1861, would be the recipient of a small fortune, which he enlarged through his investments in city real estate. Just as William H. Crocker controlled much of San Francisco's banking activity, Phelan was equally influential in real estate. Phelan loved San Francisco, for in a small degree, he *owned* it.[3] He also had ambitions for it. He studied the history and architecture of Rome, Paris, and London with the thought of replicating their most desirable features in San Francisco. His friend, Gertrude Atherton, perhaps best described him as a man with "a broad and charitable outlook, and while one side of his mind was intellectual, with a great love of literature and particularly poetry, the other was shrewd, far-seeing, financial." He never married, and when Atherton suggested that it was because "his one true love was California," he smiled and did not deny it.[4]

Phelan's instincts were urban, and when he experienced the outdoors, as he often did at his manicured Montalvo estate, his idea of roughing it was to have a barbeque complete with chilled wine, linen, and uniformed servants.[5] In essence, Phelan's compass point in life was anthropocentric: he truly appreciated the creations of human beings. Nature, when present in his life, was a mere backdrop for music, poetry, sculpture, grand architecture, and art. The night before the San Francisco earthquake Phelan enjoyed the performance of Enrique Caruso as Don José in the opera *Carmen*.[6]

In contrast, on that fateful day Muir was in the Arizona desert examining deposits of petrified wood.[7] He was never at home in the city. He was comfortable in his well-tended orchards across the bay in Martinez but never more at ease than in the pinnacles of the High Sierra. Muir was more biocentric, focused on the creations of nature. Kevin Starr, the historian of California, summarized the two men: "For Phelan, California was the splash of baroque fountains in a sun-drenched plaza. For Muir and his fellow Sierra Club members, . . . California was a trek though the High Country."[8]

The two men looked on the Hetch Hetchy Valley through the prism of their beliefs, and what they saw was quite different. From their appositional points of view spring many of the arguments the two sides advanced. At issue was the meaning of progress and whether the United States had reached the point where it might wish to preserve landscapes of special beauty, even if that might inconvenience material needs. For Phelan that point had not been reached. For him, a great dam symbolized human determination and ingenuity, an edifice that would enhance nature and yet serve the human needs of his city. Muir did not believe that humans could enhance the beauty of nature, at least not a mountain sanctuary. In a sense, the debate was the opening salvo of a century's worth of conflict over "the highest and best use" of natural areas.

During the early phases of this opening salvo San Francisco had the edge. Even before Muir formed the Sierra Cub in 1892, the city cast covetous eyes toward the valley. Mayor Phelan applied for a permit in late 1902, but did not get it. The city learned some lessons. Patience would be necessary. The supervisors found that there were other water options. Leaders also discovered that momentous events can influence decisions. The earthquake and fire, while disastrous in itself, created sympathy for the stricken city. San Francisco also recognized that power resided in Washington and that views modify with political change. From 1903 to 1914 the city of San Francisco and the Sierra Club changed their arguments very little, but how Washington received those arguments did matter. Two secretaries of the interior favored the city's position, two leaned toward the Sierra Club view. Washington political concerns, as well as changing attitudes toward the national parks and natural resources, played a crucial part in the outcome of this fight. Both sides would learn, if they did not know already, that in such natural resource issues, the political party and personalities who controlled Washington made a difference.

Mayor Phelan made the first step when he applied to the Department of the Interior on July 29, 1901, for reservoir rights on the Lake Eleanor site and in the Hetch Hetchy Valley.[9] Although this was a beginning, it was also the culmination of an unfolding interest in the valley. That interest is best represented by looking at the clouded career of William Hammond Hall, an engineer intimately involved with both Hetch Hetchy Valley and San Francisco water politics. In May of 1878 the California legislature appointed Hall as state engineer, essentially to continue the work of Josiah Whitney. One of his tasks was to collect data on water sources that could serve municipal purposes. By 1882 Hall had completed a survey of Lake Eleanor, located to the west of the Hetch Hetchy Valley and at a considerably higher elevation.

FIGURE 4. Mayor and senator James Phelan attended the dedication of the 1923 dam and gave a speech. He was heavily invested in the Hetch Hetchy water and power system from 1900 until his death in 1930. *Courtesy of the Bancroft Library.*

Eleanor Creek was a small tributary that entered the Tuolumne River west of the valley. Also in that year, J. P. Dart, a Sonora engineer, drew up maps for a Hetch Hetchy water project. Further downriver, W. G. Long and Joseph Hampton, both from the town of Sonora, filed water rights with the idea of constructing ditches to bring water to San Francisco. Interest in water transfer to the bay city was on the rise. However, these projects had as much chance at success as miners finding the mother lode. When presented to investors, the great cost involved in building a private water system to serve the city caused them to cease their scribbles and calculations and seek profit elsewhere.[10]

By 1885 the governor granted Hammond Hall leave to join John Wesley Powell's Irrigation Survey of the Western States. In the course of this work, Luther Wagoner, Hall's assistant, surveyed the Lake Eleanor site and its outlet stream. However, Wagoner passed over the attractive Hetch Hetchy Valley. According to Hall's later recollection, Powell would not approve such a spectacular valley for water storage "until necessity should call for its use, and that time, he considered, was then in the very distant future."[11] There is little reason to doubt Hall's account, but one wonders whether Powell had viewed a painting or a photograph of Hetch Hetchy or whether he was aware that shortly thereafter the valley would become part of Yosemite National Park. For whatever reason, Powell went on record that his U.S. Geological Survey would have no part of any plan to submerge the valley for irrigation or a water source.

However, Powell's reservations did not influence the government engineers for long. In 1891 John Henry Quinton inspected the valley for the Geological Survey. Aesthetic considerations no longer seemed paramount, and Quinton displayed no reluctance in recommending the valley as a proper location for a dam. However, the city—torn with corruption and dissension—was not prepared to take action. Even when Phelan won the mayorship, his first priority was to reform an outdated city charter, not find water. When he had accomplished his municipal reform through what became known as the Phelan Charter, or more properly, the 1900 Charter, he turned his attention to public works. With political power consolidated in the mayor's office, Phelan selected C. E. Grunsky as the city engineer. Grunsky, a respected civil engineer who would go on to become president of the American Society of Civil Engineers, studied every water alternative within 150 miles of the city. Over a two-year period, his assistants examined water sources from the Feather River in the northern Sierra to the Tuolumne River in the southern Sierra. He made his recommendation on January 23, 1901, when he sent a letter to the Board of Public Works recommending that the city develop the

Tuolumne River watershed, with dams at the Lake Eleanor and Hetch Hetchy Valley sites. If the city failed to attain a federal permit to do so, Grunsky hoped that federal officials would deny any pending homestead claims. Although other options were not totally discounted, Mayor Phelan and San Francisco would follow Grunsky's recommendation. Once the city made its choice, it hung onto that choice with the determination and patience of a wolf after its prey.

To assist with the city's application, Phelan hired Joseph B. Lippincott to conduct the necessary surveys. This action was suspect, for Lippincott was an employee of the U.S. Geological Survey. He should not have done private consulting work, nor should Phelan have hired him. But Lippincott seemed to have no qualms about selling his services to the highest bidder. He had the advantage, of course, of *appearing* to be an impartial government employee simply doing an irrigation survey. Just a few years later, Lippincott would be severely criticized for using his position with the Bureau of Reclamation to further the interests of the city of Los Angeles in acquiring water rights on the Owens River.[12] The situation in San Francisco was, if not identical, similar.

Phelan and Grunsky, on the other hand, can be faulted for secretly filing water rights under the mayor's name rather than that of the city of San Francisco. Phelan argued, however, that the covert action was necessary since "there were no provisions in the regulations issued by the Department of Interior under which a municipality could file."[13] We do know that Phelan filed, not out of any hope for personal gain, but rather for the benefit of his cherished city. He pondered the possibility that the Spring Valley Water Company might somehow gain senior water rights in Hetch Hetchy if the survey became public information. Such suspicions were well founded. The company had already frustrated San Francisco's efforts to establish a municipal water supply when it got wind of the city's intent to buy the Calaveras dam site (in Alameda County). Spring Valley outbid the supervisors and nipped the city's water aspirations in the bud. City pretext notwithstanding, secrecy in the world of water claims and water rights was every bit as legitimate as in that of land purchases.

In time, John Muir would identify Phelan as the true enemy of the Hetch Hetchy Valley. His activities, his determination, and his political acumen were traits that, in Muir's view, would mark him as the man at the heart of this "black job." Marsden Manson, soon to become city engineer, and Warren Olney, a charter member of the Sierra Club turned "Benedict Arnold," were "only hired promoters." "Never was a scheme more truly a one man scheme than this."[14] Phelan became the object of much of Muir's contemptuous and

FIGURE 5. The famed naturalist and writer John Muir was the spiritual and political leader of the Hetch Hetchy fight. This photograph, taken in 1907, may have been shot in the valley by Herbert Gleason. *Courtesy of the Bancroft Library.*

occasionally heated accusations. In the natural world, Muir might characterize him as a snake, slithering about in a devious and plotting way. In truth, Phelan was an ambitious leader, a compelling speaker, and a man of civic responsibility, convinced that a municipally owned Hetch Hetchy system would resolve the city's water woes.

Significantly, in his recommendations City Engineer Grunsky suggested alternatives to submerging Hetch Hetchy Valley. In light of subsequent events, his recommendations must not be forgotten. Although Grunsky preferred to dam the valley, he believed that "either Lake Eleanor or Hetch Hetchy would be sufficient for the needs of San Francisco." For the foreseeable future—for Grunsky, a century—he believed that a combination of water sources from a free-flowing Tuolumne River, a dam at Lake Eleanor, storage rights on Cherry Creek, and the significant watershed resources of the Spring Valley Water Company would be sufficient to meet the city's needs as well as that of the participating Bay Area communities.[15] The Hetch Hetchy Valley need not be touched. The valley could remain in abeyance, to be developed—if his predictions proved accurate—sometime around the year 2000.

In the end, neither Grunsky, Phelan, nor the supervisors embraced this option. They remained firmly committed to a dam in the Hetch Hetchy Valley, no matter what the financial realities or the political consequences. No person could fully explain this commitment, but perhaps San Francisco mayor Edward Taylor came closest when he suggested to a Commonwealth Club audience in 1909 that the Hetch Hetchy Valley seemed to work a magical spell on those who encountered it. The dam site and valley tended to "completely hypnotize every civil engineer that sees it, and to render him forever after incapable of a rational consideration of the larger problem of public policy relating to it."[16]

With the Right-of-Way Act of 1901 in place and Phelan's water rights filed, it was time to take action. But San Francisco had to clear one more hurdle. The city had to gain permission from Secretary of the Interior Ethan Hitchcock. On the surface city leaders were confident, particularly because no significant interest group objected. Yet certain aspects of the project might complicate the secretary's decision. Above all, the city wished to invade a national park. Furthermore, there was no love lost between city leaders and the U.S. Army, administrator of the park. The San Franciscans appeared to have a certain aloofness and sense of superiority, which annoyed those officers and enlisted men sworn to uphold the park regulations. Friction became public when, in 1896, Colonel S. B. M. (Samuel Baldwin Marks) Young ordered the detention of a camping party of prominent San

Francisco lawyers, businessmen, and aspiring politicians. In defiance of park regulations and Colonel Young's instructions, the San Franciscans carried loaded pistols and rifles on an outing near Hetch Hetchy. When the army patrol intercepted them, the soldiers faced a belligerent group that resisted arrest. When they attempted to ride off, Lance Corporal James F. Keilty grabbed the lead horse's reins and then disarmed the riders. He then led them on a three-day march to Wawona, where they received a severe dressing down by Colonel Young, giving notice that class privilege would not be extended to Yosemite National Park. Aside from the humiliation, the San Franciscans' horse trip was over, ruined, they believed, by an overzealous officer. The outraged dignitaries returned to the city and then complained to California senator George Perkins.

The senator sent a long letter to both the secretary of the interior and the secretary of war, representing the San Franciscans as gentlemen who occupy "high positions in this commercial community" and "are all influential men in their respective spheres." Lieutenant Colonel Young's action had been "hasty, ill-considered, and very reprehensible." Perkins called for disciplinary action, but what he got was an unrepentant military officer who refused to be browbeaten by a senator or his San Francisco friends. Eventually the whole incident was forgotten, and Young continued to enforce regulations in an egalitarian fashion.[17] The incident was merely the opening round of a lukewarm relationship between the city and the park that would continue until 1930.

In early January 1903 the city application for water rights in the Hetch Hetchy Valley and at Lake Eleanor lay on Secretary Hitchcock's desk. It did not take him long to make a decision. On January 20 he informed the General Land Office of his intent, and on January 30 he officially denied the city's request. He based his decision on the fact that the reservoirs were within Yosemite National Park. There was also a technical question whether some of Phelan's claim lay on private land, over which the secretary had no authority. The city immediately appealed, and to alleviate the question of Phelan's rights, the former mayor transmitted to the city his "right, title and interest" in both Hetch Hetchy Valley and Lake Eleanor.[18]

City Attorney Franklin K. Lane spearheaded the appeal process and journeyed to Washington in April to present the city's case in person. Lane argued that Hetch Hetchy water would go to the "highest possible use," but he primarily addressed Hitchcock's national park concerns. He maintained that the proposed dam would transform the meadow into a "highly attractive feature of the mountains." Furthermore, a bridge at the dam site would make the area north of the Tuolumne River accessible for recreation.[19]

Lane's aesthetic argument that the dam and reservoir would enhance the overall beauty of the park would be proffered often in the next decade. In time it would win over other interior secretaries and Congress, but Hitchcock remained unconvinced. The dam would be a violation of Yosemite National Park, which he would not sanction. Hitchcock denied the appeal on December 22, 1903, and soon after wrote President Theodore Roosevelt that "if natural scenic attractions of the grade and character of Lake Eleanor and Hetch Hetchy Valley are not of the class which the law commands the Secretary to preserve and retain in their natural condition, it would seem difficult to find any in the Park that are unless it be the Yosemite Valley itself."[20]

Secretary Hitchcock's decision boiled down to a legal issue of whether, under the Yosemite National Park Act of 1890, he was sworn to preserve the park in a natural state, or if the Right-of-Way Act of 1901 authorized—some said required—the secretary to grant water development for beneficial purposes. The secretary found his role as park protector more legally sound, as well as more appealing. No doubt Roosevelt found such an argument attractive, but he also had ears for Gifford Pinchot, his chief forester and confidant, who would soon use his influence on behalf of the city. Meanwhile, soundly rebuffed by Secretary Hitchcock, Franklin Lane returned to San Francisco to consult with Phelan and the new mayor, Eugene Schmitz. For the first time, but not the last, San Francisco realized that the invasion of a national park would not come easily, and when it did, it would be overlaid with frustrating conditions.

The first chapter in the Hetch Hetchy fight thus ended in the city's defeat. However, Phelan would not abandon his dream. Many of the city reformers were surprised and disappointed when the mayor did not seek a fourth term. For reasons he never recorded, he walked away from the office, satisfied that he had brought honest government to his city. His biographers suggest that stress, brought on by those who did not share his dream for "San Francisco the beautiful," may have influenced his decision to retreat to Montalvo.[21] Certainly other factors weighed on him, perhaps even his Hetch Hetchy plan, which now lay moribund. His withdrawal, of course, was short-lived, and soon his latent political ambitions and his commitment to Hetch Hetchy made him the unofficial ambassador for the city. Independently wealthy, he often traveled to Washington to lobby and fight for the Hetch Hetchy project. Later, as senator from California (1915–1921) he continued his support. Of all the persons subsumed by the Hetch Hetchy Valley fight, it was Phelan—who lived until 1930—who had the greatest longevity.

Into the office vacated by Phelan came Eugene Schmitz, a handsome theater musician and occasional band director. Charming in his way, he had few principles and considerable political obligations. One debt was to Abe Ruef, San Francisco's most notorious political boss. Ruef had orchestrated the musician's election in both 1903 and 1905, pulling the political strings to lead him to office. Relegated to the sidelines during the Phelan years, Ruef was anxious to get back into the game of graft, which he played with increasing boldness, especially after Schmitz's reelection. Certainly one of the most lucrative plays could be found in setting water rates. But even more profitable could be selling supervisors a new water plan. With the Hetch Hetchy system mired in controversy and with little reason to believe that Secretary Hitchcock would reverse his decision, the stage was set for new water ideas. In early January 1906 Ruef entered into an informal agreement with William S. Tevis, the president and major stockholder of the Bay Cities Water Company. The company controlled major water rights in what was known as the Blue Lakes Region, as well as claims on the headwaters of the South Fork of the American River and the North Fork of the Cosumnes River. If Ruef could convince Mayor Schmitz, his maestro of mischief, and the city supervisors to adopt the Bay Cities plan for a municipally owned water system, Ruef's cut would be approximately $1 million to be distributed to the mayor and cooperative supervisors, with the considerable residue to be kept for himself.[22]

Aside from the sordid nature of San Francisco politics, the city was determined to divorce itself from the Spring Valley Water Company. Furthermore, the Charter of 1900 mandated a municipally owned water system. To make recommendations, Mayor Schmitz appointed a committee composed of Jennings J. Phillips, James L. Gallagher, Charles Boxton, James F. Kelly, and Edward Walsh. Their report dismantled the Hetch Hetchy scheme by noting that the senior water rights claims of the Turlock and Modesto irrigation districts, never adequately addressed by Phelan, would make Tuolumne River water unavailable. Furthermore, "the utilitarian uses [in a national park] of a city is one [*sic*] that the Federal Government is not likely to tolerate." If the city persuaded the government to grant a license, it might soon be persuaded to revoke it. The committee believed that Congress would never take action on the matter, and as long as other sites were available, no secretary of the interior would grant the Hetch Hetchy Valley to San Francisco. Playing off Secretary Hitchcock's decision, the committee believed that insurmountable barriers existed with the Tuolumne River site and that the city ought to seek other options.[23]

This report resulted from a Board of Supervisors resolution on January 29, 1906, just three weeks after the board assumed office. The new board resolved to waste no more time or money on Hetch Hetchy, but rather to invite water project proposals from other sources. They received 14, but soon winnowed the choices down to 5. The supervisors preferred the proposal of the Bay Cities Water Company, but to give some legitimacy to the final decision, they appointed an "Advisory Engineering Board." Charles E. Marx, a respected professor of civil engineering at Stanford University, headed the short-lived board. Marx and the other engineers agreed that before recommending one plan, they had to study all of the proposals. The supervisors disagreed. They were in a hurry, especially after the April 18 earthquake and fire destroyed the heart of San Francisco. When the supervisors announced that they could not wait, the engineering board resigned.[24]

With the pendulum swinging its way, the Bay Cities Water Company pushed its advantage. Edwin Duryea, the company's chief engineer, appeared before the Commonwealth Club, the most prestigious forum in San Francisco. Duryea took the offensive by exposing some of the rather loose Hetch Hetchy facts presented by Phelan and Grunsky. The Tuolumne River watershed was not 1,500 square miles, but 536. The Blue Lakes area (South Fork of the American River and North Fork of the Cosumnes River), although smaller in size (396 square miles), received a greater mean annual rainfall. Whereas the mean in the Tuolumne watershed was 36 inches, that of Blue Lakes was 59.7 inches. The audience was invited to do the math: "The dependable stream-flow or water-producing capability of the Bay Cities Water Company's Catchment-area is in average years over one and two-thirds as great as that of the Tuolumne." Duryea noted that the water quality of all the Sierra sources was of unusual purity. Blue Lakes was no different, particularly since outside of a few campers in the summer, virtually no one lived within the catchment area. He believed that with stream flow and reservoir storage, the Blue Lakes system could provide the city with over 317 million gallons per day, even during dry years. Such a water supply would last until the year 1995 or 2040, depending on growth predictions. Finally, the Tuolumne River project would not be "free," as some proponents argued. On the contrary, the difficult access, the dams, the ditches and tunnels, the railroad, and roads would all require large outlays of money. However, the Blue Lakes system's proposed price, Duryea reminded his audience, would include "the transfer to the City not only of lands and water-rights, but also of all the completed structures necessary to develop a water-supply one-half greater than Mr. Grunsky's proposed Tuolumne supply."[25]

Engineer Duryea made his presentation in support of an official offer that William Tevis, president of the Bay Cities Water Company, had made to the San Francisco Board of Supervisors. In a draft dated April 9, 1906, the company proposed to sell for $10.5 million all of its improvements and water rights. Tevis judged that the distribution system to convey the water to San Francisco would "not exceed $27,500,000, and might be constructed for considerably less." Thus the total cost would be $38 million. Grunsky had estimated the cost of Hetch Hetchy at $39.5 million; thus the city would save $1.5 million by adopting the Bay Cities Water Company proposal.[26]

What was conspicuously missing from William Tevis's proposal was a discussion of power production. At a time when an electric revolution was under way and hydroelectricity development was occurring up and down the Sierra Nevada, neither Tevis nor Duryea broached the subject, aside from one sentence in which Duryea stated that 30,000 horsepower of energy could be developed.

Duryea declined to discuss hydropower because three founders of the Pacific Gas and Electric Company had already developed powerhouses, or at least acquired the rights to do so. The development of hydropower in the Mokelumne-Cosumnes river system came with the rather romantic involvement of a French prince with the unlikely name of André Poniatowski. Prince André, as he was commonly called, gave evidence of the cosmopolitan nature of California and the international interest in mining. After learning the mining business in South Africa, the prince was drawn to California. However, he soon abandoned mining for a new source of wealth: power production. Like many European noblemen, the prince had a fine lineage but little money. He resolved that problem when he married Elizabeth Sperry, the sister-in-law of William H. Crocker. With his marriage he also consummated his power interests. After a series of mergers, André had controlling interest in the Consolidated Light and Power Company, which served the San Francisco Peninsula. In 1903 the prince decided to return with his family to France. He sold his power company interests to John Martin, Eugene de Sabla, and Frank G. Drum, three of the founders of the Pacific Gas and Electric Company, incorporated in 1905. These power transactions did not directly involve William Tevis or Lloyd Tevis, his landholding father, but clearly William understood that while San Francisco might have the water of the Bay Cities Water Company, the power would be reserved for Pacific Gas and Electric.[27]

The Bay Cities Water Company proposal had some merit, although historians have dismissed it as nothing but a graft scheme that would have saddled the city with essentially worthless claims.[28] However, in 1900 the city professed to be interested in a water supply, not a power source. If that was true, then it might have been wise to pursue Tevis's proposal. Surely, a number of the company's water rights claims were suspect, but the city could easily have written a contract that withheld payment until Bay Cities made good on its promises. Furthermore, the Hetch Hetchy water system was not without its payoffs to private water interests.

The problem with Bay Cities, therefore, was not so much with the message as with the messenger. By the time that Duryea argued before the Commonwealth Club, the San Francisco graft trials were underway. A court convicted Ruef and sentenced him to serve time in San Quentin prison. Mayor Schmitz also faced corruption charges but would escape the penitentiary. City government, in disarray, dropped the Bay Cities water plan like a hot potato. In a hostile atmosphere, the Bay Cities proposal became a casualty and the city shelved it forever. Furthermore, the episode stoked popular prejudice against offerings by private enterprise to resolve the city's water problem. In a different political climate, Bay Cities might have received a serious hearing, and the waters of the Tuolumne River might still flow gently through the Hetch Hetchy Valley.

FOR SAN FRANCISCANS everything changed on April 18, 1906. At 5:12 A.M. the city shook for 28 seconds, the quake measuring 7.9 to 8.3 on the Richter scale by modern analysis. Buildings fell, power lines separated, gas lines broke, and of course, water mains ruptured. Many people never woke. Some of those who did perhaps wished they had not. With overturned kerosene lamps and broken gas mains, fire engulfed four square miles in the city center for four days. Thousands of refugees fled the city, mainly to the East Bay. Others camped out in Golden Gate Park or anywhere that was free of fire and falling walls. Enrico Caruso left the city in haste, never to return. John Muir, fossicking in the Arizona desert, returned to his much damaged home. His daughter, Wanda, wrote that all the houses in the surrounding Alhambra Valley were a wreck. The Muir home had lost all five chimneys, and as Wanda put it, "I never saw such a smash in my life. The whole house has to be rebuilt. What shall I do?"[29] In the passion of the moment, Wanda exaggerated. Muir headed for home, and aside from the chimneys, the cracks and fissures in the old house were repairable.

The earthquake was a natural disaster that few San Franciscans could blame on anyone, save those who viewed the event as God's wrath unleashed on a sinful, pleasure-loving city. However, the fire was another matter. As the flames spread across the city, firefighters and citizens called for water. They found little. The violent shaking had broken many of the main lines. The fire hydrants of the Spring Valley Water Company delivered only a trickle. It is unlikely that even if water had been available, it could have made a significant difference, but the helplessness of being unable to do anything but watch the flames leaping ever higher was galling. Residents looked for answers. It was only natural to blame the water company for such a fiery tragedy. In truth, it mattered not a whit if the broken water mains were privately or publicly owned, yet people looked for a scapegoat and the company became an easy target.

Almost immediately the city supervisors authorized a Committee on the Reconstruction of San Francisco, with a subcommittee appointed to investigate the water supply and fire protection. Its eight members were mostly engineers, headed by C. H. McKinstry of the Army Corps of Engineers. Both Charles Marx of Stanford and Marsden Manson, the city engineer, sat on the committee. The supervisors asked for not only recommendations to avoid the recent catastrophe but also ideas regarding future water sources. The committee refused to address the second question, reasoning that the answer would involve an extensive engineering survey. In regard to the fire recommendations, however, the members concluded that "the protection against fires afforded by the system of Spring Valley Water Company was inadequate" for the crisis it faced. The committee felt it imperative that the city control its own water supply. Having gently chastised the water company, the engineers paradoxically declared that they had no criticism of the workmanship or the materials provided by Spring Valley. What they did recommend was a "special fire-protection system and the acquisition of a municipal system . . . as quickly as possible."[30]

Surely this report ruffled the feathers of William Bourn, chairman of the board of the Spring Valley Water Company. The committee gave no explanation of the "special fire-protection system" and seemed not to admit that breaks in the water mains were inevitable. The real miracle was that Hermann Schussler's Crystal Springs Dam, located directly on the San Andreas fault, withstood the shake with no damage. If the dam had broken, the failure would have caused more deaths, and the city would have suffered with little water for many months.

In spite of the engineering soundness of the Spring Valley system, it was inevitable that "The Great Earthquake and Fire" would give added momentum for the city to acquire a municipal system. Most residents would agree with one engineer who wrote Robert Underwood Johnson that "our late earthquake and fire has opened the eyes to the fact that we have the poorest water-system of any city in the United States."[31] Of course, the water supply committee, under the guise of objectivity, favored municipal ownership and was not hesitant to use the earthquake and fire to further its purpose. With the catalyst of the great fire, the city and Spring Valley once again began the dance of acquisition, and the city—once again—turned its eyes east to the great Hetch Hetchy Valley. Appearing before the supervisors in April 1908, A. H. Payton, president of Spring Valley, responded to the city's need for five million more gallons daily. The city had two choices: buy the system for $32 million ("the best terms available") or let the company do the job through expansion, particularly in Alameda County and the Calaveras dam site.[32]

In the meantime the San Francisco earthquake and fire created national sympathy for the prostrate city. Offers of assistance from individuals, cities, and governments appeared daily, in a lavish display of generosity. Soon enough this compassion for San Francisco would play out in the halls of Congress. Who could, after all, deny the struggling metropolis its desired water supply on hearing Key Pittman's emotional appeal to his fellow senators?

> I happened to be in the City of San Francisco in 1906 when that great catastrophe came. I was there when the earth was shaken by that terrible quake. I saw the fire break out all over that city and sweep from one end of it to another. I saw homes crumble and swept away as the wind might blow a stack of cards. I saw women with children at their breasts filling those parks.
>
> I saw the hillside covered by homeless people. I saw such suffering as I never expect to see again, and I know that a lot of it was caused by reason of the inefficient system of water that was being supplied to the people of San Francisco. I know it was largely due to the greed of that water monopoly in its efforts to spend as little as possible and to grasp just as much as possible, and I never want to see such a condition again exist in any city.[33]

Although Pittman's description was overdrawn, City Engineer Marsden Manson knew that the time was right to reintroduce the Hetch Hetchy idea. He looked to U.S. Chief Forester Gifford Pinchot for help. Both had been dismayed when Secretary Hitchcock refused to give San Francisco Lake Eleanor and the Hetch Hetchy Valley. Manson wrote Pinchot soliciting his help. Pinchot responded, noting that he was pleased to hear that the earthquake "had damaged neither your activity nor courage." In regard to water, the politically powerful forester sincerely hoped that the city would "be able to make provision for a water supply from the Yosemite National Park, which will be equal to any in the world. I will stand ready to render any assistance which lies in my power." Naturally, Manson was thrilled that the city had a vigorous voice in Washington. In November 1906 Secretary Hitchcock, Manson and Lane's nemesis, departed from the Interior Department. His replacement happened to be a good friend of Pinchot. With a somewhat transparent meaning, Pinchot wrote Manson that Hitchcock's successor, James Garfield, might be more amenable to their plan for the Hetch Hetchy region. He could not forecast the actions of the new secretary, but "my advice to you is to assume that his attitude will be favorable, and to make the necessary preparations to set the case before him." Pinchot had heard rumors that San Francisco had given up on the "Lake Eleanor plan" and decided to go elsewhere for Sierra water. However, if the Hetch Hetchy idea was still open, he wrote, "by all means go ahead with the idea of getting it."[34]

With such encouragement Manson, Phelan, and the Board of Supervisors resurrected the moribund Hetch Hetchy plan. In July 1907 Secretary of the Interior James R. Garfield traveled to San Francisco, and Mayor Edward Taylor arranged a meeting to push reconsideration of Hetch Hetchy. He then appointed Manson and Phelan to present the San Francisco case. Taylor realized that the two were the best informed and that they, in the words of Phelan's biographers, "offered the best blend of technical understanding, humanistic sensitivity, and political aplomb."[35] They met on the evening of July 24. In addition to the San Francisco spokespersons, P. J. Hazen and A. C. Boyle represented the Turlock and Modesto irrigation districts. The meeting focused on irrigation rights, and Manson promised that the two districts' senior water rights would be fully protected. Garfield said little.[36] Yet the San Francisco delegation successfully planted an idea that would be amplified, resulting in the Garfield grant of May 1908. This important grant authorized the city to proceed with its Hetch Hetchy plans.

No Sierra Club members were present at Garfield's July 1907 evening meeting. However, the gathering did signal San Francisco's intentions, and it acted as a catalyst for Muir and the club. Certainly the club knew of the city's interest, but until the middle of 1907 there were no protest resolutions, and club member's concerns were voiced in private. The 1903 Hitchcock decision seemed to foster a false sense of security. The club forgot that what one secretary of the interior could give, another could take away. Muir and others believed they could depend on the government to protect its national park. We have no record of the conversations between Roosevelt and Muir from their memorable four-day outing in 1903, but the two could not stay away from politics for long. No doubt Roosevelt expressed sympathy for the preservation of Hetch Hetchy, and Muir would have taken such sympathy as a declaration for the valley. At the time, the club was tiny, and devoted to convincing the California legislature to rescind Yosemite Valley to the federal government, and the U.S. Congress, in turn, to accept its lost possession. Convincing two separate jurisdictional bodies was not easy, and the effort consumed the energies of what was essentially a Bay Area club. Exhausted by its successful 1905 effort to return Yosemite Valley to federal control, the club seemed passive in the face of a threatening storm over Hetch Hetchy.

But now the city's renewed activity provided that "firebell in the night" for Muir and William Colby, the industrious young secretary of the Sierra Club. John Muir used his renown to inform such clubs as the Mazamas, the Appalachian Club, and walking clubs in Chicago of the city's designs on the deep, spectacular valley. He also wrote Theodore Roosevelt. The naturalist expressed his concern that Yosemite Park be saved from commercialism "other than the roads, hotels, etc. required to make its wonders and blessings available." He was, of course, most concerned with Hetch Hetchy. He ranted against such men as Phelan and Manson, stating that their arguments "all show forth the proud sort of confidence that comes of a good sound substantial irrefragable ignorance." Ever since Congress established Yosemite Park, its friends had had to defend it. What could be done? He hoped that president could help. In his appeal, Muir unveiled a striking analogy. "The first forest reserve was in Eden," he reminded the president, "and though its boundaries were drawn by the Lord, and angels sent to guard it, even the most moderate reservation was attacked." The president responded from his Oyster Bay home, stating that he would do all in his power to protect Hetch Hetchy, but "so far everyone has been for it and I have been in the disagreeable position of seeming to interfere with the development of the State [of California] for the sake of keeping a valley, which apparently hardly anyone wanted to have kept, under national control." Roosevelt ended his letter in a reminiscent mood, wishing that he could see Muir in person, and lamenting that he was not again with him "camping out under the great sequoias, or in the snow under the silver firs!"[37]

Embedded in the president's words was a message. Muir must build a constituency if he was to have any success with the San Francisco juggernaut. To do so would require more than a local effort. To rally national interest, Muir called on his old friend Robert Underwood Johnson. He and Colby also appealed to J. Horace McFarland, a great friend of the national parks and executive director of the American Civic Association. Both men would play crucial roles in the coming fight.

McFarland was no newcomer to battling engineers intent on progress at the expense of scenery. Niagara Falls, the greatest natural wonder in the eastern United States, had been under attack for many years. Frederick Law Olmsted had deplored the tacky commercialism that seemed to sprout up overnight, knowing that such intrusions detracted from the fall's picturesque and sublime beauty.[38] At the turn of the century, McFarland faced a more aggressive and skilled enemy: professional civil and electrical engineers who looked on the falling water with covetous eyes. These "scientific wizards" were the heroes of the day who, with the general support of the American people, were intent on absolute control of natural resources for what they saw as the advancement of civilization. With the nation in the midst of an energy revolution, such electrical engineers as Charles Proteus Steinmetz seemed larger than life. Steinmetz, in an article entitled "Mobilizing Niagara to Aid Civilization," suggested that it was time to harness the immense potential of the falls through diversion and hydropower development. For those such as McFarland who were profoundly concerned with the loss of a scenic attraction and tourism, Steinmetz had a bizarre solution. Why not let the falls "run dry" during the week and then "turn them back on" for the tourists on the weekend?[39]

At the time Muir and Colby first contacted McFarland, he was in the thick of his struggle to save Niagara Falls. Fortunately he found allies in Lord James Bryce, the British ambassador to the United States and author of the classic work *The American Commonwealth,* and Theodore Burton, chairman of the House Committee on Rivers and Harbors. The Burton bill, which became law in June 1906, combined with an international treaty signed in 1909 by Bryce for Canada and Secretary of State Elihu Root, ensured that Niagara's waters would flow for people's appreciation, not for their factories.[40]

McFarland found many similarities between his fight for Niagara and what was developing on the West Coast. He would become an indefatigable ally of the Hetch Hetchy Valley: a man who, like Robert Underwood Johnson, had the ear of prominent senators and even presidents. He often operated out of the Cosmos Club, founded by John Wesley Powell in 1878 and boasting such members as Theodore Roosevelt, William Howard Taft, and many men of science and power. The club was, as Wallace Stegner has written, "the closest thing to a social headquarters for Washington's intellectual elite."[41] From such a gathering place McFarland could reach an audience unavailable elsewhere. His biographer suggested, however, that his dedication to Hetch Hetchy may have been inspired not by the refinements of the Cosmos Club but by a sign from nature. At his home in Pennsylvania a noisy woodpecker would wake him every night at 3:20 A.M. Rather than a mere annoyance, the flicker was, in McFarland's view, "trying to sound an alarm; to rouse him from his reveries of building his garden to the real dangers threatening the one God had made centuries before in California's Yosemite."[42] McFarland responded.

The Sierra Club took its first formal action on August 30, 1907, when the board passed a lengthy resolution opposing the use of Hetch Hetchy as a reservoir site. The resolution stressed the beauty of the valley, its tourist potential once the government constructed a road, the inviolate nature of a national park, and the fact that San Francisco could find pure and abundant water elsewhere.[43] A few months later a letter signed by Muir, Colby, Joseph N. LeConte, William Badè, and E. T. Parsons, all club leaders, urged the membership to write President Roosevelt and Secretary of the Interior Garfield.[44]

WITH THE CLUB and his friends alerted to the danger, in the fall of 1907 John Muir's thoughts turned often to the Hetch Hetchy. Twelve years had passed since he was last in the valley. It was time to reacquaint himself with the object of his interests; its sheer granite cliffs, lush meadows, and particularly the deep-running Tuolumne River. Tueeulala Falls would be nearly dry and Wapama Falls greatly diminished, but the river would cast its spell on the mountaineer. He rejoiced in its running water. Hal Crimmel, writing about Muir in Alaska, remarked that "water simply enchanted Muir his entire life."[45] When camped along the Tuolumne River, he remarked once that "for my part, I should like to stay here all winter or all my life or even all eternity."[46] Perhaps it was the auditory factor, transforming itself from a deafening roar in the Grand Canyon of the Tuolumne and Rancheria Creek, to the murmuring, gurgling hush as it flowed through the valley. The river, with its slow pace and multitude of currents, put Muir in a reflective mood. Here was a river that had its beginning high in his mountains and its ending at sea: one a birthing, the other closure, or death. Yet he knew the river was immortal. He could count on the Tuolumne to flow, continuing to charts its course, as well as his. A reservoir, of course, would bring premature death to a river.

The chance to revisit Hetch Hetchy with a cherished friend was an opportunity not to be missed. William Keith accompanied him. An artist by nature and a Scotsman by birth, Keith met Muir in 1872 in Yosemite. The two found much in common, not only in their backgrounds but in their interests as well. Although Keith was basically apolitical, he found in the mountains the subject of a lifetime of work. The two were born in the same year, and they both "inherited the love of nature implicit in the Gaelic temperament."[47] They spent many days together in the mountains, and Muir would often visit Keith in his San Francisco studio. A visit with Keith could make the city tolerable. There was no one Muir would rather have accompany him on his revisit to the Hetch Hetchy Valley. In early October Muir wrote his friend, "I'm glad you have decided to take a breath of mountain air. It's what we both need."[48]

At 69, neither man was as vigorous as he once was. They rented horses, and at Croker's Station they hired a man to look after the stock. Even before they left, Muir exulted in being back in the mountains. He could appreciate any of nature's seasons, but fall suited him. In a letter Muir penned to his daughter, he exclaimed that "the glory of the woods hereabouts is now in the color of the flowering dogwood—glorious masses of red & purple & yellow beneath the pines & firs." To Colby he wrote that "Keith is enjoying himself in this bracing air, and I hope to get rid of my cough in blessed and bothersome Hetch Hetchy."[49]

The next day they arrived in the "blessed and bothersome" valley. We have little detail of their stay, but undoubtedly they wandered about the valley, Keith with his easel and Muir with his notebook. In the shorter October days, they would have conversed at length around a comforting campfire. Of course when it came to Hetch Hetchy Valley, Muir always had a political agenda. Later he wrote his Los Angeles friend Theodore Lukens, who had spent time with Muir in Hetch Hetchy in 1895, that "Keith and I returned from a two week visit to Hetch Hetchy a short time ago. Keith was charmed with it—said it was finer & more beautiful & picturesque than Yosemite [Valley]. He made 38 sketches." To *Land of Sunshine* magazine editor Charles Lummis, another Los Angeles friend, he briefly described the trip and then added: "If successful in getting large enough appropriations for a good wagon road into the Valley & a trail up the big Tuolumne cañon the salvation of the glorious Hetch Hetchy will be made sure for then it will be *seen* & known by countless thousands making effective lying [by San Francisco] impossible."[50] With his call for development, Muir, contrary to the idea of saving the vanishing wilderness, devoutly wished for the wilderness to vanish.

As the year 1907 drew to a close, no one could doubt the determination of San Francisco to secure Hetch Hetchy. Phelan, Manson, and city officials had set aside all the other options. In the nation's capital, the triumvirate of Roosevelt, Pinchot, and Garfield leaned strongly to giving San Francisco what it wanted. Of course all three agreed that the Hetch Hetchy Valley should remain undefiled until a future time when it might be needed. Feeling more than confident, the San Franciscans descended on Washington to press Garfield to grant the city use of the Hetch Hetchy Valley. They expected very little resistance. In that assumption they were wrong. Muir, revitalized by his recent encounter with the valley, was prepared to fight.

CHAPTER 4

Two Views of One Valley

"Short-haired women and long-haired men."

CITY ENGINEER MARSDEN MANSON

"Despoiling gain-seekers and mischief-makers
of every degree from Satan to Senators."

JOHN MUIR

In early 1908, after five years of frustration, the San Francisco team had every reason to be confident. Their view of the use of Hetch Hetchy Valley loomed large. They had the support of Gifford Pinchot, and they expected that President Roosevelt would follow the lead of his chief forester. Secretary Garfield, although he had not made his final decision, leaned toward San Francisco. These three national figures made a formidable alliance. Roosevelt was the strongest politically, although the weakest link philosophically. Yet San Francisco anticipated that the personal influence of Roosevelt's friend, University of California president Benjamin Ide Wheeler, as well as the pleas of Phelan and Manson, would trump the arguments of Muir and Robert Underwood Johnson. Psychologically, the city could capitalize on the nationwide sympathy for a community devastated by earthquake and fire. Politicians would find it difficult to oppose the city's desire for a pure, reliable water supply, whether it was in a national park or not. The leaders of the Modesto and Turlock irrigation districts, although still opposed to the city's water ambitions, were willing to talk as long as San Francisco acknowledged and guaranteed their senior water rights on the Tuolumne River. All the forces seemed aligned for victory.

Most encouraging, there was no significant opposition. True enough, Muir and Colby had urged Sierra Club members in August 1907 to protest San Francisco's plans. However, the members had not responded as the directors might have wished. It was soon evident that the club was hopelessly divided, floundering in dissension and disarray. The Hetch Hetchy issue revealed a deep schism, somewhat along geographical lines. Many of the San Francisco members favored municipal use of the valley, while those at greater distance shared Muir's view. The club would not speak with one voice on this issue. Hermann Schussler, the skilled Spring Valley Water Company engineer, and Marsden Manson, both members, fell on the side of the dam, and they influenced others toward rebellion.

For many, it was a painful decision. The case of Warren Olney bordered on a Greek tragedy. He was one of the original organizers of the club in 1892, and it was in his San Francisco law office that the founders gathered to draw up the articles of incorporation. He was an avid mountaineer and a great admirer and friend of Muir. However, Olney had moved to the East Bay and served as mayor of Oakland. From that perspective the line between public good and national park protection became blurred. He struggled with his advocacy of a regional water supply and his love for the Hetch Hetchy Valley, which he had visited and enjoyed. Reluctantly he stepped across the line to the San Francisco side. In the years to follow, Olney and Manson would argue furiously on behalf of the city. For Olney, in particular, the fight was not without personal anguish. Later his daughter revealed that "after twenty years of pioneering service and close friendships" in the club, his resignation was a bitter pill. So bitter, his daughter recalled, that "the Hetch Hetchy project was never afterward a permissible topic of conversation in our household."[1]

On the national level, Gifford Pinchot's position was crucial. As earlier mentioned, the influential forester argued in favor of San Francisco's case, but he professed to still have an open mind. He promised John Muir in September 1907 that the two would meet, since Pinchot seemed unaware that Hetch Hetchy comprised a significant part of Yosemite National Park.[2] The meeting did not take place, and Pinchot avoided visiting the valley or directly engaging with any of Hetch Hetchy's defenders.

However, Muir and Pinchot were not so alienated as the standard narration would suggest.[3] The friendship between the two men worked well in the late nineteenth century, but deteriorated in the twentieth. According to Pinchot's biographer, Char Miller, the two first met in the Adirondacks in 1882. They spent a few very enjoyable days hiking together, Pinchot assuming the role of student to his well-known mentor. Muir found Pinchot an intelligent, adventurous young man, at home in the woods, unfazed by inconvenience, and like Muir, oddly eager to "relish, not run, from a rainstorm." Pinchot was Muir's kind of person, and the two hiked often and conversed even more.[4]

Their friendship became strained in 1897, when the two men served on the National Forest Commission, charged with evaluating President Cleveland's controversial forest reserve proclamations. Muir sided with the commission chair, Charles Sprague Sargent, who believed the reserves ought to be inviolate. Pinchot was not so sure. Undisturbed by their lack of harmony, Muir and Pinchot still often walked together, arguing, agreeing, and ultimately confiding in each other. Yet they did not share a common view regarding the intrinsic value of living species. Their differences were apparent in a trifling incident when the two men camped together in the Grand Canyon. On that escapade Pinchot recalled that when they "came across a tarantula he wouldn't let me kill it. He said it had as much right there as we did."[5] Muir defended the life of one of the most feared, least loved noxious spiders known to man. Pinchot could see no reason why the tarantula should live, and was quite prepared to crush its life away. It had no real benefit to man that Pinchot could identify. Muir intervened because of his respect for all life, sentient or not. Every living being had its place in the broad scheme of things, and "lord man" should not needlessly intervene or attempt to divine what that purpose might be. Pinchot was inclined to exert his power through a well-placed boot stomp. Muir urged, indeed demanded, human restrain. If we substitute a valley for the tarantula, we can understand what was at the heart of their disagreement over Hetch Hetchy.

Supposedly, the final break occurred in September 1897 in a Seattle hotel lobby, when Muir marched up to the forester with a newspaper in hand which quoted Pinchot as saying that sheep grazing in forest reserves did little harm. When Pinchot admitted to the quote, Muir, according to biographer Linnie Marsh Wolfe, gave his ultimatum: "Then if that is the case, I don't want anything more to do with you."[6] However, Wolfe may have taken some journalistic liberties. Three months after the Seattle encounter, a letter from Muir expressed his delight in hearing from Pinchot, writing that he would "be glad to hear how you succeed in your forest plans."[7]

Sheep may not have spoiled the relationship, but Hetch Hetchy did. Their differences had much to do with politics. Pinchot believed in the art of the possible, always weighing the needs of human beings in his equations. The forester understood that in the world of Washington he would do his best to respect the natural world but if human needs intervened, there was no question where he stood. Muir had the luxury of being both critic and advocate.

Since he had no political ambitions and was beholden to no person or group, he could speak, write, and defend the natural world with an eloquence and purity seldom heard. He surely understood politics, and he was not indifferent to human needs. He always maintained that San Francisco deserved, indeed must have, an adequate water system. But whether it was water or sheep, he believed in what we might call "zoning for nature." There were places where cities might find water and herders could fatten their sheep, but in that determination the natural world must have a voice.

Muir provided that voice in no uncertain terms. When it came to Hetch Hetchy, Muir had a difficult time maintaining civility with those who disagreed with him. He felt too deeply and too passionately to be a detached diplomat for the Hetch Hetchy Valley. This inflexibility sometimes worked against him. In retrospect, Muir may have lost an opportunity to save his valley when he failed to explore fully Pinchot's, Roosevelt's, and even Manson's view that San Francisco would be willing to develop first the Lake Eleanor site, leaving the Hetch Hetchy Valley for a later date, perhaps 50 years hence. Muir might have vigorously probed this option, seeking compromise and gaining a lengthy time extension for the valley, and perhaps preserving it for all time.[8]

Much of Pinchot's position flowed from his belief that the use of Hetch Hetchy for a water supply represented the greatest good for the greatest number of people. Beyond that dictum, so firmly embraced by many of the California Progressives, he was convinced that a very innocent Muir and his friends were blind to the public power issue. It was Warren Olney, the most influential club rebel, who portrayed the Sierra Club board and his former friends, as well meaning but naive dupes. In defending the valley, they had allied the club with the Spring Valley Water Company, which, of course, opposed the development of the Hetch Hetchy for very different reasons. Even more serious, the club had unwittingly become the mouthpiece for the Pacific Gas and Electric Company, incorporated in 1905. Although Muir would often rant about the "capitalists" ransacking the national park, there was much irony in the charge. Many observers, and certainly the San Francisco press, believed the Sierra Club was not defending the valley but rather *private monopoly*. In opposing public power, the club seemed not only to protect exploitive capitalist enterprise but to care little for human life. Writing to President Roosevelt, Robert Underwood Johnson felt it necessary to clarify his position. "Human life is more sacred than scenery," he assured the president, "and *if it were the only alternative,* sooner than see the city or California go without an abundant supply of good water I would cheerfully dam up the Yosemite itself!"[9]

With discord as its password and its supposed disdain for human life part of its image, the small coterie of nature lovers were in trouble. The condition of the Sierra Club paralleled the physical health of its leader, which was not good. Caught up in three bouts of the "grippe," similar to influenza, in early March 1908, Muir wrote Johnson that the disease "came near making an end of me." To his friend Theodore Lukens he confessed the illness had caused "anxiety & sorrow & sickness" making him "barren and useless."[10] In early 1908 the survival of the Hetch Hetchy Valley was in serious doubt.

As the year wore on, both sides continued their lobbying efforts in Washington. Muir urged Roosevelt to give the city only Lake Eleanor. He asked that "the special pleaders," the "graft lawyers," and those involved with the "miserable dollarisk squabbles" be turned aside. He also had harsh words for Phelan and Benjamin Ide Wheeler, Roosevelt's friend and the head of the University of California. Roosevelt responded, suggesting a softer tone. Muir ought "not run down those men too much. Benjamin Ide Wheeler and other good fellows are among them." In the meantime Roosevelt assured Muir that he was trying to follow a path that would spare the valley for at least a generation. But then he added, "I must see that San Francisco has an adequate water supply."[11]

Roosevelt's concluding sentence suggested an ominous note. Shortly after the letter exchange, Secretary Garfield received an eight-page petition, dated May 7, 1908, from City Engineer Manson. He formally asked that the secretary reopen San Francisco's application for water rights. He apologized for sending copies of maps of location, for the original maps had been destroyed in the fire following the earthquake. The most significant section of the application contained Manson's duplicitous treatment of the Hetch Hetchy Valley and Lake Eleanor. Convinced that the city could not afford to develop Lake Eleanor without the Hetch Hetchy site, Manson urged that the government grant the use of both sites. The city engineer promised to develop Lake Eleanor first, but San Francisco would determine the time frame. In further clarification, stipulation 3 read: "The City and County of San Francisco will develop the Lake Eleanor site to its full capacity before beginning the development of the Hetch Hetchy site, and the development of the latter will be begun only when the needs of the City and County of San Francisco and adjacent cities, which may join with it in obtaining a common water supply, may require such further development."[12] The two statements in concert clearly declared that San Francisco would not invade the Hetch Hetchy Valley until such time as the city had a demonstrated need for more water. As Pinchot suggested, that time would probably not come for at least fifty years. Yet if approved, the petition would put the power of decision in the hands of San Francisco. The city would decide when it was time to submerge the Hetch Hetchy Valley, and neither the public nor the government need be consulted. Approval of the petition required great faith that the city would spare the Hetch Hetchy Valley until need required otherwise.

Secretary Garfield approved the San Francisco petition only four days after receiving it, suggesting that he had previously consulted with Roosevelt, and perhaps Pinchot, and had fully intended to grant San Francisco its wish ever since the evening meeting in June 1907. Armed with an opinion from the attorney general, the secretary believed that he had the full authority to grant water development in a national park. The grant would provide the greatest benefit to the greatest number—a catchall axiom of the Progressive reformers. Furthermore, it authorized development of a hydropower system to be constructed and owned by the city. For transportation San Francisco would build and maintain a highway from the Big Oak Flat Road. Picking up on Manson's postponed use of the valley, Garfield declared his appreciation that the city would not invade Hetch Hetchy until its needs exceeded the water supplied by Lake Eleanor. He was encouraged that the city "has expressed a willingness to regard the public interest in the Hetch Hetchy Valley and defer its use as long as possible." To his critics he described how the lush meadow will become "a lake bordered by vertical granite walls." The meadow, somewhat unusable and mosquito infested, will become "a lake of rare beauty."[13]

The Garfield grant represented chapter two of the controversy, a clear victory for San Francisco. It contained the arguments put forth by the city, and in a sense it was merely the fruition of the ideas of Phelan and Manson, harmonized with the social philosophy of Pinchot and Roosevelt. There was, unfortunately, a misplaced faith in the metropolis. In signing the grant, Garfield relinquished the federal government's power to determine the construction schedule of the proposed Hetch Hetchy dam based on a water needs assessment. The Garfield grant was so "loose" that, in classic political style, the secretary gave the city what it wanted but then assured himself and his critics that San Francisco would hold to the letter and spirit of the agreement. Muir, of course, was disappointed in the decision, but he expected it and could take some comfort in the fact that Hetch Hetchy was not immediately on the chopping block. He wrote Johnson that "the Valley will escape damming most likely in our day at least," and he bolstered Colby's sagging spirit with the thought that the valley was safe for many years to come.[14]

No sooner had the San Francisco position been firmly established than it began to unravel. Almost immediately Congress got involved. On May 16, a scant five days after the Garfield grant announcement, Representative Julius

Kahn of California introduced House Joint Resolution 184 "for the purpose of exchanging lands between the city and the federal government."[15] The resolution would require congressional hearings. For the first time, the Hetch Hetchy issue would enter the halls of Congress, with a hearing scheduled for December 1908. With their executive branch options nearly exhausted, the valley's defenders viewed hundreds of senators and congressmen as a new opportunity to promote their view toward national parks in general and the Hetch Hetchy Valley in particular.

The importance of a congressional opportunity became doubly apparent when the nature lovers were cut out of participation in the 1908 Conference of Governors regarding conservation. Pinchot controlled the invitations, and with the feeble excuse that the White House room lacked space, he declined to invite Muir, Robert Underwood Johnson, and others who happened to differ with his opinions. Only J. Horace McFarland, representing the American Civic Association, received an invitation. Nevertheless, swallowing its pride, the Sierra Club board sent a greeting to the governors and President Roosevelt, reminding the assembled group that "our country has a wealth of natural beauty which is far beyond the power of human hands to create or restore, but not beyond their power to destroy." Mindful that the governors might be swayed by economic considerations, the board noted that Europe's "beauty is worth $550,000,000 annually." America affords the world's newest "pleasure-grounds," and "tourists of wealth and fashion . . . are flocking in constantly increasing numbers to the Cordilleran system of mountains on our Western coast." The governors should make every effort to tap into this tourist bonanza, while securing for "coming generations the benefit of our scenic resources."[16] Although the Sierra Club board did not specifically mention Hetch Hetchy, the message was clear: The valley should be spared for its *tourist* value.

Although Pinchot denied Muir's participation, the famed naturalist made his views known from afar. When J. Horace McFarland invited his friend to make a statement on the national parks and Hetch Hetchy, Muir used the opportunity to make a passionate, although not necessarily reasoned, pronouncement:

> But however abundantly supplied from legitimate sources, every national park is besieged by thieves and robbers and beggars with all sorts of plans and pleas for possession of some coveted treasure of water, timber, pasture, rights of way, etc. Nothing dollarable is safe, however guarded. Thus the Yosemite Park, the beauty glory of California and the Nation, Nature's own mountain wonderland, has been attacked by

> spoilers ever since it was established, and this strife I suppose, must go on as part of the eternal battle between right and wrong. At present the San Francisco Board of Supervisors and certain monopolizing capitalists are trying to get the Government's permission to dam and destroy Hetch Hetchy, the Tuolumne Yosemite Valley, for a reservoir, simply that comparatively private gain may be made out of universal public loss.[17]

To those who argued for the replacement beauty of a reservoir, Muir responded: "As well may damming New York's Central Park would enhance its beauty!"

McFarland made use of Muir's emotional language in his address to the assembled governors, noting that the "Hetch Hetchy Valley of the Yosemite region belongs to all America and not to San Francisco alone."[18] McFarland's brief but dynamic talk could not turn the tide. Robert Underwood Johnson, who made an unofficial appearance at the conference, wrote Muir that he did his best but that any opponents of San Francisco were "all considered sentimentalists." In truth, Muir's inability to see the issue in nuanced terms diminished his arguments with such a politically sophisticated gathering.

In the meantime, Sierra Club executive secretary William Colby dedicated his efforts to a strategy for the December congressional hearings. The Kahn bill must be vigorously opposed, he argued, for if passed the act would "have a strong tendency to perfect the [Garfield] grant." Looking to the future, the savvy Colby believed that if Congress passed the act supporting San Francisco's intent, it would be difficult for a succeeding interior secretary to go against Congress by revoking the Garfield grant. Already Colby was looking to a time when James Garfield and his friend Pinchot would be relieved of power. In the meantime, however, the Sierra Club director was committed to presenting their case to Congress, primarily through "a little pamphlet in an attractive a form as possible." It would contain some of Muir's writing on the valley, as well as the best photographs available. The idea would be to present with words and photographs just what the nation would lose should Congress side with the city.[19]

Although the exact numbers and dates of the pamphlets produced by the Sierra Club and, later, by the Society for the Preservation of National Parks are difficult to determine, the first printed circular appeared shortly after Garfield's grant, with the cumbersome title "Mr. John Muir's Reply to a Letter Received from Hon. James R. Garfield in Relation to the Destructive Hetch Hetchy Scheme." In the years to follow these pamphlets, with such titles as "Save the Hetch Hetchy Valley," "Prevent the Destruction of the

Yosemite Park," and "Let Everyone Help to Save the Famous Hetch Hetchy Valley," reached numerous people on the eastern, middle, and western parts of the nation.[20] They were particularly effective in the Boston area, where Allen Chamberlain saw that every member of the Appalachian Mountain Club received one. J. Horace McFarland provided both labor and money to ensure that members of the Cosmos Club and his American Civic Association were informed. Robert Underwood Johnson mailed brochures to his influential friends and acquaintances.

With Congressman Frank Mondell of Wyoming presiding, the House Committee on the Public Lands opened hearings on December 16, 1908. Marsden Manson testified for the city, explaining in detail that San Francisco wished to exchange some 500 acres of Hetch Hetchy Valley land for city-owned land that fell within Yosemite National Park. No sooner had hearings begun than Mondell adjourned until January 12, to observe the holiday break.

When the hearing resumed, Manson made the rather startling statement that the city would not be interested in Lake Eleanor without assurances that it could *immediately* develop the Hetch Hetchy Valley. Manson argued that Lake Eleanor would never supply the city's long-term water needs over the next fifty years, given a projected population of 1.5 million to 2 million. Besides, as city engineer he could not recommend such an expensive construction project to the voters of San Francisco without assurance of a reservoir in the Hetch Hetchy Valley.[21] San Francisco, therefore, did not want Lake Eleanor if it could not have Hetch Hetchy. Manson's position was a shocking development, for it announced that the city intended to renege on its previous promises. The announcement also contradicted the position of the threesome of Garfield, Pinchot, and Roosevelt, yet none made the effort to call Manson to the carpet. It was, quite frankly, a disgraceful abandonment of an agreed-upon position that offered a sensible compromise. To John Muir it came as no surprise. Earlier he had written Johnson that the San Francisco leaders were "going to Congress to make a desperate effort to get what they want, as I knew they would. Viz, permission to dam H. H. at once without reference to Lake Eleanor & Cherry Creek. For it is Electric power they want, not water. Of course we must fight 'em & I think we'll beat 'em."[22]

Muir was right on the city's motives as well as the outcome of the hearing. Most members of the Public Lands Committee received more than a hundred letters of protest, written at the urging of Muir and Sierra Club board members Joseph N. LeConte, Edward T. Parsons, and William Badè. Their circular "To All Lovers of Nature and Scenery" stressed the natural beauty of the valley and the idea that a national park should not be violated needlessly.[23] The letter writers advocated preserving the valley as a natural place for the American people to enjoy and appreciate. Through such letters and Sierra Club pamphlets, congressmen who had never heard of the Hetch Hetchy Valley now became informed. The isolated valley was gaining name recognition.[24]

The most damaging testimony for San Francisco, however, came not from Sierra Club members but from Edward J. McCutcheon, testifying as attorney for the Spring Valley Water Company. Exacerbating the long-standing tension between the company and the city, McCutcheon maintained that there was no immediate water crisis. Therefore, it made sense to slow down. He reminded the congressmen that "Hetch Hetchy Valley is not going to run away." The congressmen were treated to a heated argument between McCutcheon and City Engineer Manson when the Spring Valley attorney raised the subject of a buyout. For years the city and company had haggled over the purchase price. Actually the spread was not great, with the city willing to pay $28.5 million and the company seeking about $32 million, but even this small chasm could not be bridged. In such disagreements the city should have exercised its right of eminent domain, allowing an arbiter to determine the final price of purchase. McCutcheon maintained that Spring Valley was quite willing to enter into the process, yet the city was not. McCutcheon suggested the reason:

> The city of San Francisco is afraid to leave to an impartial tribune the determination of the value of this property, and it comes to you, the Representatives in the Congress of the United States, and says, "Our hands are tied. We are shackled by this monopoly; we are afraid to seek the aid of the courts of San Francisco for the purpose of acquiring this property, and we ask you to give us this big stick in order that we may wield it over the Spring Valley company and make it come to our terms."[25]

Manson and A. P. Giannini responded that the city believed it could come to an agreement without going to the courts, but considering the many years of acrimony, such an excuse hardly seemed plausible.[26] San Francisco emerged from the House hearings somewhat battered, and without the victory Phelan and Manson anticipated.

The next month Hetch Hetchy moved to the Senate chambers for hearings on a companion resolution. This time the valley defenders were much better prepared. Robert Underwood Johnson, Alden Sampson, and William Badè for the Sierra Club; Edmund Whitman for the Appalachian Club;

Harriet Monroe of Chicago; and J. Horace McFarland for the American Civic Association, all testified. Muir, in poor health as he often was in winter, could not attend. However, he volunteered to pay all the travel and lodging expenses for Badè and Johnson to appear.[27]

Senators Francis Newlands and Reed Smoot both attempted to dilute testimony by connecting the interests of the valley defenders with the Spring Valley Water Company. Senator Smoot asked Whitman directly about his connection with Spring Valley. The Boston lawyer replied, "I have no pecuniary interest in this matter whatsoever." Senator Newlands, who had once served as attorney for Spring Valley, asked Johnson the same question and perhaps got more than he bargained for: "I have not the faintest interest in the Spring Valley Water Company, and not one of us who appear here have any interest in it," Johnson declared, adding, "as representatives of the public we refuse to be put out of court on the ground that we happen to travel along the same lines as the argument of the Spring Valley Water Company."[28] Johnson was annoyed at the question, but no matter what Johnson, Muir, Colby, or anyone else could say, the suspected connection and possible collusion haunted them throughout the struggle.

This formidable group of witnesses stressed the beauty of the Hetch Hetchy Valley as well as the sanctity of the national parks. Johnson proposed that the nation had four "great natural features": the Grand Canyon, Yosemite Valley, Niagara Falls, and the Hetch Hetchy Valley. He named the Yellowstone country as fifth. Perhaps the most eloquent testimony came from poet Harriet Monroe, who had spent a week in the summer of 1908 in the valley. In a moving tribute, she called it a garden of paradise, unequaled. Aside from Muir, Monroe would become the most eloquent voice for protection from inundation. Whitman, who had also visited the valley, praised the scenery and particularly the Grand Canyon of the Tuolumne, the magnificent 20-mile river canyon between Tuolumne Meadows (8,600 feet) and Hetch Hetchy (3,800 feet). He envisioned building a scenic road through the canyon so that others might enjoy it, an idea that would horrify Sierra Club members today.[29]

Alden Sampson, a great admirer of Muir and an avid mountain climber, favored total development of the valley, to include hotels, "roads and trails in every direction with small hostelries where people will be put up overnight." Millions of people, he believed, should be able to enjoy the scenery. As for San Francisco's water needs, Sampson had a bizarre idea. He asked the senators, who had topographical maps before them, to note a number of high-country lakes, elevated above Hetch Hetchy and within the Tuolumne River watershed, that, he suggested, could be dammed. The stored water, when released,

FIGURE 6. This serene prospect gave ample evidence of the opportunity for contemplation in the Upper Hetch Hetchy Valley. *Courtesy of the Bancroft Library.*

would skirt the valley by aqueduct on its way to San Francisco. Considering the loss of at least 10 natural lakes and landscape that would be devastated by road construction and maintenance, it was fortunate that no senator took the suggestion seriously.[30] There is no evidence that Samson presented his idea to the Sierra Club board or anyone else.

As the hearing concluded, Edmund Whitman suggested that the senators pass a resolution asking President Taft to appoint a special commission to investigate the Hetch Hetchy issue. Both the senators and the witnesses agreed this would be an excellent idea, although Johnson made a plea for professional diversity. He did not want a commission comprised solely of engineers. The Senate committee was enthusiastic for the commission idea, as it allowed them to withdraw the bill from consideration and set it aside for further study.[31] They could postpone a decision on a contentious issue.

It seemed that Muir's prediction that they would "beat 'em" was indeed coming true. At the opening of 1908 Phelan and the San Francisco forces felt confident that they had won the battle. A year later, the *San Francisco Call* voiced disgust that debate in Congress had turned to "a lot of talk about 'babbling brooks' and crystal pools." Nothing much could be done until "the New England nature lovers exhaust their vocabularies."[32] Both the *Call* and city officials realized it would not be easy. The congressional hearings revealed a determined opponent with nationwide connections that San Francisco had not anticipated. Furthermore, in March 1909 William Howard Taft assumed the presidency of the United States. *Collier's Weekly,* supporting the claims of San Francisco, believed the worst of Taft. There would be no grace period for the new president. In regard to Hetch Hetchy the weekly magazine announced that this "simple soul" would be "the most gullible President, in regard to his associates, since Grant left the battlefield for the White House."[33] Furthermore, Secretary of the Interior James Garfield resigned, and Taft's new secretary, whoever he might be, would likely be less sympathetic to San Francisco's hopes. What a difference less than a year had made.

The city had made a serious tactical error when it supported the introduction of the Kahn Resolution (H.J.R. 184) for the purpose of sanctioning land exchanges between the federal government and the city. The attorney general's office had determined that Garfield had full authority to make the grant to San Francisco, exclusive of Congress. There was no compelling reason for the city to let the issue slip from the executive branch of government. The necessary land exchanges could have been negotiated between the Department of the Interior and the San Francisco city attorney. If the courts determined it was necessary for confirmation by Congress, the exchange could have been presented as a fait accompli, requiring a mere voice vote by the two houses. Instead, Kahn's resolution stirred a sleeping giant, one of such size and energy that it would rock the city back to a defensive position.

For San Francisco, more bad news would come. There is a principle in Western water law that "if you don't use it, you lose it." San Francisco was about to find out the truth of that phrase. Joseph Lippincott of the U.S. Geological Survey, City Engineer Carl Grunsky, and Mayor Phelan had perfected water rights claims at both Lake Eleanor and the Hetch Hetchy Valley in 1901. However, while the city made surveys and some improvements in the valley, validating its claim, it ignored Lake Eleanor and its outlet. Although Grunsky valued the water potential, even suggesting that Lake Eleanor water alone could meet the city's needs for many years, he focused on the Hetch Hetchy Valley site. As a consequence Grunsky and his staff made no improvements at Lake Eleanor, in effect abandoning the city's water right claim.

Such an abandonment would have been inconsequential had no one been watching. Unfortunately for San Francisco, William Hammond Hall, an engineer of some ability and an opportunist who sought profit, noticed and prepared to pounce. July of 1902 found Hall camped at Lake Eleanor tacking up claim notices. He was involved in a classic case of claim jumping, but this time it was water, not minerals. His attorney had wired him in Sonora saying, "Abandonment complete and positive, in my opinion." Creating the stories of later Hollywood movies, Hall outmaneuvered not only San Francisco but other water seekers. According to him, two Los Angeles engineers also planned to post Lake Eleanor, but they were three days late. "Old Kibbe," a wizened California prospector who lived near the lake, recalled that after seeing Hall's claim notices, the two engineers "hit the trail for Sonora, cussin' like blazes."[34]

With his claim secure, Hall became a middleman for Frank Drum, Eugene de Sabla and John Martin, three principal founders of the Pacific Gas and Electric Company. While fronting for "his clients," Hall approached James Phelan twice in 1907, knowing that Lake Eleanor was bound to be an integral part of any Hetch Hetchy development. He was, as one might expect, met with hostility. Phelan could not quite believe that, through neglect, the city had lost the water rights he had worked to file. When finally convinced, Phelan demanded to know who Hall's principals were, assuming he represented either the Spring Valley Water Company or the Bay Cities Water Company. Hall revealed nothing, simply stating that his clients would entertain a purchase price of $300,000 from the city for the water rights and improvements of the Sierra Ditch and Water Company. Phelan showed him the door.[35]

Rebuffed by the city, Hall sought out another purchaser. He settled on John Hays Hammond, a Californian born in San Francisco in 1855 but a mining engineer who left the state with the close of hydraulic mining in 1884. Hammond gained a worldwide reputation and a considerable purse through his activities in South Africa, Russia, and Mexico. He believed devoutly in his profession, convinced that engineers and their "machine civilization" utilized the physical and chemical resources of the world for human advancement. He was, in a sense, a supreme representative of a new faith, based on capitalism, technological knowledge, and a commitment to change. He believed that such places as Hetch Hetchy represented "portions of the globe which for eons had remained comparatively barren and useless" but now might be "transformed into a blessing to man."[36]

After some months of haggling, Hammond secretly purchased the Sierra Ditch and Water Company for Hall's $300,000 price tag, hoping to sell his water rights to the city while developing the hydropower. He retained William Hammond Hall to deal with San Francisco or any other interested parties. At this point Hall became deeply enmeshed in city politics, and in time he would be tarred as the bandit who stole Lake Eleanor and then held it for ransom for eight years.[37]

The city, embarrassed by its oversight, vilified Hall yet reluctantly had to negotiate. As the year 1909 progressed, the Lake Eleanor–Cherry Creek property continued to appreciate. Manson called for an impartial appraisal of the property, and Charles D. Marx, a Stanford engineering professor, and consulting engineer J. G. Galloway produced a report, dated August 28, 1909. The two engineers suggested a total payment of $150,000 for Hall's Lake Eleanor landholdings and water rights.[38] Manson found the price too high. However, in April 1910 the city finally capitulated, purchasing the Lake Eleanor and Eleanor Creek property for $400,000 and the Cherry Creek water rights and property for $600,000. The Cherry Creek (often called a river) water rights actually represented a very significant new purchase, for Phelan's 1901 water filing had never included the creek.[39]

This Lake Eleanor water rights adventure, unfortunately, tied San Francisco more tightly to the Hetch Hetchy project. Considering the time, effort, and money invested by the city, it would be difficult to extricate itself. In acquiring all privately owned lands and water rights in the Hetch Hetchy Valley, along the Tuolumne River, and the Hall-Hammond property, the city had spent $1,915,000.[40] This sum represented a serious commitment of San Francisco taxpayers' money. With such an investment it would be foolhardy to walk away and in effect abandon the project, much as logic might indicate such a course. San Francisco could not consider other water options, but must simply plow ahead, hoping for victory but never admitting defeat.

NOT ONLY was the San Francisco municipal government rocked back on its heels by Hall's unsavory water dealings, but in March of 1909 it also had to face a new administration. The sympathetic Washington triumvirate of Roosevelt-Pinchot-Garfield was no more. Gifford Pinchot would stay on to serve in the new administration, but not for long. Theodore Roosevelt was off to Africa, and Phelan must have felt a twinge of concern as he wired him, "BEST WISHES. GOOD VOYAGE. GOOD LUCK. SAFE RETURN."[41] Replacing Roosevelt and Garfield would be President William Howard Taft and Secretary of the Interior Richard Ballinger, neither of whom had announced, publicly or privately, a willingness to champion the city's cause.

Sierra Club members were equally unsure of the new president and his interior secretary. Muir sought out information, while Robert Underwood Johnson made contacts in Washington. Johnson was pleased that Ballinger intended to visit Hetch Hetchy Valley in the summer "in hopes of becoming better informed."[42] Furthermore, the new president announced to Johnson that he intended to come west, and Yosemite Valley would be on his agenda. With his political matchmaking skills, Johnson insisted that the new president and his secretary meet Muir in San Francisco. Taft responded: "Tell Muir to join me on my arrival in San Francisco and I'll be delighted to have him on the Yosemite trip." A few days later Muir received a formal invitation from Frank Carpenter, the president's secretary, asking if Muir could join Taft on October 6. Muir, of course, was delighted.[43]

In early October President Taft arrived in San Francisco for his tour of Yosemite Valley. Muir accompanied him, and the two got along quite famously. Writing to his friends Katharine and Marion Hooker of Pasadena, Muir described the president as "the merriest man I ever saw & he makes all his company merry. The birds & squirrels & deer were half charmed & frightened." Perhaps it was his rotund physical size that alerted the wildlife, but Taft was still quite athletic. He walked the four-mile trail down from Inspiration Point, chatting all the way, while, according to the *San Francisco Call* reporter, setting a pace that left the others "calling for help." Some of the chat, of course, involved Hetch Hetchy, and Taft laughingly suggested that since the valley was so far from any center of population, it might just as well be used commercially. The jest translated into a headline in the *Call*, "President Taft Chaffs Muir on Sentimentality," with subheadlines that the president listened "with good nature to Naturalist's Frantic Shriek of 'Sacrilege.'" The newspaper assured its readers that San Francisco's Hetch Hetchy project had nothing to fear. But the reporter who filed the story was not invited to the noonday meal. There Galen Clark, then 97 and one of the great Yosemite pioneers and protector of the Big Trees, joined Taft and Muir, as did Major William Forsyth, Yosemite's superintendent. All three men questioned San Francisco's motives and intentions, which was bound to make an impression on the president.[44]

After spending three days with Muir, President Taft suggested that his interior secretary might also have Muir as a companion-guide, since Ballinger would arrive the next week for an inspection trip of the Hetch Hetchy Valley. With the blessings of the president, Muir greeted Richard Ballinger at the El Portal railway terminal. Off they rode to the Hetch Hetchy Valley, and although no detailed account remains, Muir believed Ballinger was won over by the trip. On their return he wrote Colby that "all seems coming our way, and the silly thieves and robbers seem at the end of their scheme." He informed Katharine Hooker that "the H. H. scheme seems doomed."[45] A letter to Ballinger, written later, revealed that the two spent time discussing a reversal of the Garfield grant and particularly the role of Pinchot, who Muir believed "seems bent on stirring up barren strife to blot and becloud the good work he has hitherto done. Whom the Gods wish to destroy they first make mad."[46]

Back in Washington, the defenders of Hetch Hetchy continued their campaign with letters inspired by a pamphlet, dated November 1909, entitled "Let Everyone Help to Save the Famous Hetch Hetchy Valley and Stop the Commercial Destruction Which Threatens Our National Parks."[47] The tract contained quotes from newspapers and magazines, photographs of the valley, and some of Muir's most powerful "sermons" stressing the folly of San Francisco's mistaken schemes.

The pamphlet represented one of the first efforts of a new organization: The Society for the Preservation of National Parks. The Hetch Hetchy issue had estranged many members of the Sierra Club, to the point that it was difficult to speak with one voice. Roosevelt had made it clear that only a groundswell of public support would save Hetch Hetchy, but how could that happen when the club could not even agree? William Colby began to think of another organization—one that would have a national voice and attract only committed members. The idea had been proposed by Edmund Whitman, the Boston attorney much dedicated to saving the valley. Muir concurred. By April of 1909 Colby had created, largely through his own initiative, the Society for the Preservation of National Parks. Muir would serve as president, and William Badè as vice president. Colby's wife, Rachel Vrooman Colby acted as secretary-treasurer.

Although Colby founded the society to defend Hetch Hetchy, its larger purpose was to protect all national parks from invasion. To ensure that the society achieved national stature, Colby called the Bay Area group the "California Branch" and then saw to it that the Advisory Council represented all areas of the country:

Allen Chamberlain	Appalachian Mountain Club (Boston)
Asahel Curtis	Mountaineers (Seattle)
Henry E. Gregory	American Scenic and Historical Preservation Society (New York)
Robert Underwood Johnson	*Century Magazine* (New York)
Harriet Monroe	Poet and founder of *Poetry* magazine (Chicago)
J. Horace McFarland	President American Civic Association (Washington)
John W. Noble	Former secretary of the interior (St. Louis)
Alden Sampson	University Club (New York)
C. H. Sholes	Mazamas Mountain Club (Portland)
Edmund Whitman	Appalachian Mountain Club (Boston)[48]

With the possible exception of John Noble, all the members would contribute their time and talent, and on occasion their money. With the officers of the Society for the Preservation of National Parks and an eager national advisory board in place, Colby wrote Pinchot a challenging message: "Let me assure you that we have only begun our fight, and we are not going to rest until we have established the principle 'that our National Parks shall be held forever inviolate.'"[49]

Colby's challenge, however, rang hollow as long as the Sierra Club was hopelessly split. Manson, Olney, Professor Alexander McAdie, William Beatty, former California governor George Pardee, and a number of others continued to denounce the club's leadership, believing that a pure water supply was worth the sacrifice of the Hetch Hetchy Valley. They provided ample ammunition for the city of San Francisco and its newspapers. The club was being torn apart. Finally on December 18, 1909, Colby sent an open letter to the membership. There had been so much bad press and questioning of the Board of Directors' position that it was time to poll the membership. Colby made it clear that the board had never opposed the development of Lake Eleanor and Cherry River for the city's use, only the violation of the Hetch Hetchy Valley. Members were asked to vote for one of two statements: (1) "I desire that the Hetch Hetchy Valley should remain intact and unaltered," or (2) "I favor the use of Hetch Hetchy Valley as a reservoir for a future water supply for San Francisco." Both sides distributed broadsides printed on club stationery. Warren Olney took full advantage, writing a seven-page statement. Among other things, he reminded the membership that it would be many years before Hetch Hetchy would be utilized, for the Garfield grant required that Lake Eleanor be developed first; thus "most of you can probably feel secure that the right to camp in Hetch Hetchy and enjoy its scenery untouched will not be taken away in your time."[50]

Colby responded for the Board of Directors and surely attacked the soft underbelly of Olney's argument when he sarcastically stated that Olney's camping statement was "entirely aside from the real question at issue. Does Mr. Olney think that you and I are so selfish that we would stand in the way of San Francisco's proposed use of Hetch Hetchy because our personal pleasure would be jeopardized?" Colby outlined the more significant issues at stake, emphasizing the violation of a national park. Evidently he was more convincing, although most members had made up their minds long before the broadsides. The final vote supported the Board of Directors' position, 589 to 161.

The opposition, however, had one more political trick to spring. The club's revised bylaws allowed that 30 members could petition for a special meeting. This they did, resulting in a meeting held on February 18, 1910, in San Francisco's Merchants' Exchange Building. The intent of the petitioners was to censure the board for its Hetch Hetchy position, but Colby was determined that this would not happen. Both sides had done their best to rally sympathetic members to attend, and Colby secured a number of proxy votes, should they be needed. The result was near chaos. Marsden Manson presented the city's case, while attorneys William Gorrill and Clay Gooding argued for sparing the valley. Each side presented a resolution, which was then hotly debated. In the end, there was so much interruption and outrage that the chairperson declared the meeting closed, with nothing accomplished save the loss of tempers. Colby frustrated the petitioners, but Muir also took an active role. Later he wrote an amusing account of his premeeting nefarious activity:

> I managed to get about 80 of our Club members out to dinner at the Poodle Dog restaurant. The long flowery table flanked with merry mountaineers looked something like a Sierra canyon so they called it, the Muir Gorge. After dinner we marched to the meeting, every face radiant, and entered the hall in a triumphant rush, fairly overwhelming the poor astonished Hetchy dammers. Next morning, the *Call* reported that I had packed the meeting, though forsoothe I had only packed seventy-seven stomachs.[51]

Quite thoroughly defeated, the "Hetchy dammers" became of little account, and the club could, with confidence, oppose the intent of the city. This victory was not without consequences, as approximately 50 members resigned in protest. Other dissenting members stayed in the club only so they could occasionally support the city's aspirations while publicly identifying themselves as Sierra Club members.

With a national organization in place and the wounds of the Sierra Club somewhat healed, Muir and Colby worked to introduce as many people as possible to the Hetch Hetchy Valley. They believed that the valley had an enchantment that could not be denied. Exposure to Herbert Gleason's fine photography of the valley was one way, but actually being *in the place* was preferable. If one could spend time there, Muir believed that one's understanding and appreciation would come through an outdoor, visceral bonding experience with Hetch Hetchy. He had taken plenty of strenuous hikes, but he didn't believe in them. Such outings were all about human egos, not nature. "Hiking is a *vile* word," he once advised Sierra Club members. "You should saunter through the Sierra."[52] His belief that *seeing* the valley in a leisurely fashion could be a transforming experience seemed borne out in the attitudes of the four secretaries of interior who, over time, determined its fate. Richard Ballinger and Walter Fisher, who spent time in the Hetch Hetchy Valley, lined up against San Francisco's desires, while James Garfield and Franklin Lane, who never visited, favored the dam and reservoir. Of course, the secretaries were predisposed to their positions, but still the Sierra Club leadership believed that actually experiencing what would be lost was important. Although Muir and Colby were at odds with Gifford Pinchot, they continually urged him to join one of the Sierra Club outings to see what was at stake. Pinchot declined, and although he spent some time in the Sierra Nevada high country, he never visited Hetch Hetchy.

Most of those who spent time in the valley did so with the Sierra Club outings in 1908 and 1909. Appalachian Club member Allen Chamberlain gained new perspective and commitment after a week in the mountain sanctuary. At the conclusion of the 1909 outing, Colby wrote McFarland that it was a great occasion, with Muir, Edmund Whitman, Allen Chamberlain, Harriet Monroe, Edward Parson, and many others coming away with renewed conviction that the valley must be spared.[53] Later J. Horace McFarland came under its spell when he accompanied the Walter Fisher party in 1911. However, perhaps the person most transformed by such a sojourn was Harriet Monroe, a Chicago poet who joined the 1908 and 1909 outings to Tuolumne Meadows and Hetch Hetchy.

Harriet Monroe was 48 years of age at the time, and just reaching the height of her creative power as a poet. In 1913, the final year of the Hetch Hetchy debate, Monroe began publishing *Poetry: A Magazine of Verse,* a particularly influential quarterly that introduced Americans to the "Imagist" poets. Her view of nature was also much influenced by long trips to China in both 1910 and 1911. Through the magazine and her anthologies she counted Ezra Pound, William Butler Yeats, and Carl Sandburg among her friends and had an acquaintance with almost every poet of her time. She loved the American West, freely admitting that "there was nothing like the wildest West to cure me of literary disappointments and other gloomy moods."[54] Writing of her 1908 visit to Hetch Hetchy, Monroe reminisced that once in the "secret valley" they walked "three level miles through flowing green grasses shoulder-high—the only human things between those granite walls, where never a hut nor a spade marred the locked inviolate wilderness." On this leisurely trip they spent three days of "enchanted wanderings—up the Rancheria Creek, back to the Little Hetch Hetchy Valley, across the river and under the cliffs; and three nights of enchanted sleep under the high pines and the stars, with the full moon mounting late over the lofty granite shoulder of 'Kolana,' and looking down serenely on the human intruders in her quiet world."[55]

Monroe had been a defender of Hetch Hetchy before this trip, but the experience of being there transformed her into a devotee. She would soon testify before the Senate Committee on Public Lands, and men like Johnson, McFarland, and Alden Sampson agreed that she was the most effective spokesperson for their cause. Her enthusiasm and her poetic descriptions won the attention, if not the hearts, of the senators. She wrote a lengthy paean to Secretary Ballinger of her love for the valley, and much as she relished the primitive place she experienced, she could not help but stress its tourist value, noting that "the beauty of Italy saves it from starvation."[56] After hearing of her political triumph, Colby asked that she "work up the scenic side of it [the valley] somewhat along the line of your admirable talk before the Senate committee."[57] In the years to follow, Harriet Monroe was the most effective spokesperson, aside from Muir, for the preservation of Hetch Hetchy. And she was quite willing to play that role, traveling from her Chicago home to Washington or California.

In her role as advisory board member to the Society for the Preservation of National Parks, Monroe wished to be included in strategy sessions. After her testimony she wrote Herbert Parsons that the fight should center on two issues: (1) the beauty of the valley and (2) the sanctity of the national parks. She was quite convinced that the club should abandon other arguments, such as one that involved sanitation. She also gave practical advice, insisting

FIGURE 7. Harriet Monroe lived in Chicago, where she wrote poetry and launched *Poetry* magazine in 1912. She loved the Hetch Hetchy Valley and wrote of its beauty and testified before Congress. *Courtesy of the University of Chicago Library.*

that Sierra Club activists should not send congressmen special delivery letters that would arrive at night. If the senders could hear the outraged, emotional remarks of some congressmen, they would understand. Later, when Taft elevated Chicagoan Walter Fisher to interior secretary, Monroe felt that he would be open to all points of view. She proved to be right.[58]

Harriet Monroe may have been exceptional in her commitment, but by

1908 Muir and Colby understood the importance of the yearly Sierra Club outing. Colby organized the first outing for the summer of 1901, following a model used by the Appalachian Club and the Mazamas of Portland. In July of that year approximately 100 Sierra Club members traveled to Yosemite Valley. The congenial group became somber when Joseph LeConte, the famed geologist, suffered a heart attack in the valley and died. Although some felt the outing should be abandoned, most agreed that LeConte passed away in a special place that he loved and that they should continue the outing.

They then hiked to Tuolumne Meadows while teamsters and their wagons hauled the 15,000 pounds of baggage, equipment, and provisions up the old Tioga Pass mining road. In Tuolumne Meadows they settled down for two weeks, day hiking, eating, occasionally swimming, and conversing. Participants experienced a close but comfortable experience with nature. The campfire was the special highlight of each day, and usually Muir, Frank Soulè of the University of California, William Keith, or naturalist C. Hart Merriam would give a talk. But just as often groups would give rather impromptu plays or all would join in a square dance, music provided by a fiddle or two. Theodore Hittell, a lawyer and historian and, at age 70, the oldest member to attend the 1901 outing, described the camaraderie and lighthearted fun of such an occasion as well as the more daunting hikes, such as up Mount Dana and back in a day.[59]

Sierra Club outing participants were of all ages and were evenly balanced between men and women. Reflecting the times, there was no racial diversity, and the level of education was very high. Whatever social barriers might exist in a more urban setting were broken down by camp life and in the shared vigor of hiking, swimming, fishing, and even mountain climbing. Bonds formed and so did commitments to protect the mountains they so enjoyed.

Not all women who cared about conservation issues traipsed around the Sierra Nevada mountains, however. The defenders of Hetch Hetchy found an effective ally in the General Federation of Women's Clubs, founded in 1890. With a membership of 800,000 women and chapters in every state, the organization was an effective group, with an affinity for conservation issues and the national parks. In a presuffrage era, it was one of the few ways in which women could gain a voice in national issues. In 1910, for instance, 233 clubs reported that they had sent letters and petitions to state and national legislators regarding forest conservation policies. These many petitions originated with the most active branch of the organization, the Forestry Committee, whose members worked tirelessly to plant trees not only for civic beautification but also to preserve pristine wooded tracts.[60]

The Forestry Committee supported Gifford Pinchot's policies, but when it came to Hetch Hetchy, they broke ranks. Historian Samuel Hays notes that women could not understand Pinchot's support of San Francisco. "Following their more basic distrust of 'commercialism' of all kinds, they vigorously opposed the reservoir. Many 'preservationists' wrote the National Conservation Association [which Pinchot controlled] to resign from a group of false prophets." By 1913 and the Raker Act debates, the influence of the General Federation of Women's Clubs (GFWC) worried Pinchot and the coterie of San Francisco supporters. William Kent, fighting for passage in the House of Representatives, contacted Pinchot to tell him that the "nature lovers" were working through the women's clubs. Harry Slattery, secretary of the National Conservation Association, scornfully remarked that the female "scenic friends" had let "their heart run away with their head."[61]

Slattery's remark merely revealed his lack of awareness. Colby had been in close contact with Jessie B. Gerald, the head of the Forestry Committee, since early 1909, soliciting support for Hetch Hetchy. When Gerald sent out 38 letters to various state "Forestry women," Colby suggested wording for the letter. In late 1909, when it was time to circulate petitions, Gerald sought help again, noting to Colby, "We do not want any amateurish thing to be sent around do we?"[62] All documentation in the *Congressional Record* related to the various Hetch Hetchy hearings reveal that the GFWC and the various lesser-known women's clubs played a significant role through letters and petitions against the damming of the Hetch Hetchy Valley.

Of course, women did not then possess the vote, which diminished their power in the eyes of many politicians. Some women used that reality to subtlety make a statement for suffrage. In response to Robert Underwood Johnson's appeal to write congressmen on the Hetch Hetchy issue, Dora Ranous responded:

> As for me—I am only a poor petticoat creature without a vote—what would my (?) senators and representatives care for my protest?—I will show the open letter to such men as I know, and test the values of "feminine influence," so highly recommended by the "Antis," and maybe I can get *them* to protest![63]

Ranous's clever strategy suggests another factor in evaluating the role of women in the Hetch Hetchy struggle. In the main, they were politically invisible, remaining in the background but working their ways, as Ranous did, with their male or female friends and evidencing the dedication of thousands of women to the cause of conservation.[64] One such women was Janet Richards, a professional lecturer who was quite willing to give a talk on

Hetch Hetchy if the Society for the Protection of National Parks would stand the cost of some lantern slides. During the battle, she gave an evening lecture in Washington to wives of senators and representatives, some of whom brought their husbands.[65] Marion Randall Parsons also dedicated herself to the cause. Married to Edward T. Parsons, aide and friend of Muir, Marion met her husband while on the Sierra Club outing of 1903. Both became heavily involved in Hetch Hetchy, and as Muir's health began to fail, Edward acted as his friend's amanuensis, often answering his correspondence. Marion, an accomplished fiction and nonfiction writer, assisted. She wrote moving letters on the beauty of the Sierra Nevada, particularly Hetch Hetchy. When Edward died suddenly in early 1914, Marion assuaged her grief by becoming John Muir's secretary, staying close by his side and assisting in the writing and editing of Muir's *Travels in Alaska*.[66]

Although Muir was fairly tough skinned, occasionally Parson and Colby found it necessary to encourage and protect their aging leader. Attacks came from the San Francisco press and the city's advocates of the Hetch Hetchy plan. The press often labeled Muir and his supporters "nature fakers," suggesting a bogus, impractical cause. Muir, who seemed above personal reproach, became feminized: a flighty, impractical person who seemed much detached from the practical needs of people, preferring to gaze at stars, reflect on the flow of running water, or describe the effervescent world of a water ouzel. Muir might engage in such activities, but such a person should not presume to enter or even comment on the practical water or power needs of cities and societies. These were male prerogatives. Phelan and the San Francisco engineers viewed Muir much as stolid New England farmers viewed Henry David Thoreau: They were both rather useless idealistic visionaries who eschewed the business of making a living, preferring to spend their days in baffling ways. Of course, Muir was, after 1880, a very practical orchardist who, true to his Scottish heritage, drove a hard bargain and was not averse to making a profit while he was at it. However, the practical Muir was a personality that few people in the Bay Area or elsewhere knew.

No one played the gender card better than City Engineer Marsden Manson, who was the chief spokesman for San Francisco's designs on the Hetch Hetchy Valley. Muir's writings, so attractive to women, annoyed Manson. Although the engineer was a member of the Sierra Club and had published in the *Sierra Club Bulletin*, he delighted in lampooning Muir's flowery writing style.[67] Muir's views were far too full of "verbal lingerie," allowing the romantic mountaineer to speak of "networks," "veils," "downy feathers," "plumes," and "embroideries."[68] Such writing by the archetypal "nature faker" obscured the practical facts that Manson believed "any sane human" could grasp. "Short-haired women and long-haired men" dominated these rabid sentimentalists, he claimed.[69] Perhaps the most flagrant attack on Muir came from the *San Francisco Call*, surely one of Manson's favorite newspapers. The fiery newspaper ran a cartoon of Muir as a plump woman with a housewife's dress, flowered hat, and a broom, sweeping back the flood of water issuing from the "Hetch Hetchy Project." One could read the iconography differently, but most would see Muir as a middle-aged female fool vainly trying to hold back the flow of progress.[70]

We do not know if Manson read Henry James, but he would agree with the novelist's 1885 concern that his "whole generation is womanized; the masculine tone is passing out of the world," ushering in a "feminine, nervous, hysterical, chattering, canting age, an age of hollow phrases and false delicacy."[71] Criticism of the city's Hetch Hetchy plans smacked of such misplaced, false delicacy, and Manson made this clear. Women, and Muir, should confine their activities to home and the bassinet. His intemperate writings probably hurt San Francisco's cause. Manson was too hot tempered, too talkative, unable to listen or engage in rational debate. He was passionate and viewed "conservation as the public application of engineering expertise."[72] He was incensed when Muir and Colby published the November 1909 pamphlet, calling it "twenty odd pages of garbled quotations and specious statement" that were utterly untrue.[73] When J. Horace McFarland accompanied Manson and Secretary of the Interior Walter Fisher to Hetch Hetchy Valley in 1911, McFarland decided that rather than contest every point with the caustic Manson, he would just let him talk until he hung himself. It was a perfect strategy.[74]

In their righteousness Muir and Manson had much in common. Manson believed that technology and engineering would create an advanced civilization. Muir believed there were limits. And much as each might vilify the other's motive, as historian Kendrick Clements has noted, "the tragic irony of the Hetch Hetchy controversy was the city and its opponents were seeking to serve the public interest as each understood it."[75] Manson was quite convinced that he was a Progressive reformer, whose passion served the people of San Francisco. Muir was no less fiery in his cause. Much like his firebrand father, Muir was uncompromising. He may have rejected his father's harsh brand of Christianity, but he did not necessarily eschew his father's fervent style. Muir simply saw divinity in the natural world, substituting it for Christianity as a guiding philosophy. He proselytized for that philosophy with unrestrained enthusiasm. Muir, according to one view, successfully took "biblical language and inverted it to proclaim the passion of attachment, not to a supernatural world but to a natural one."[76] All of this is to say that in many ways Muir was a zealot, quite determined to bring salvation to the Hetch Hetchy Valley. Like any true believer, he employed hyperbole, identifying any opponent of the valley's *deliverance* as agents of Satan, greedy capitalists, or corrupt politicians. Writing in 1908, Muir castigated not only individuals but the materialism of the times: "In these ravaging money-mad days monopolizing San Francisco capitalists are now doing their best to destroy the Yosemite Park." These "cunning drivers," he wrote, worked "in darkness like moles in a low-lying meadow." They wanted the reservoir "simply [so] that comparatively private gain may be made out of universal public loss."[77] No person of sensitivity and good moral standing could favor the damming of the valley. When it came to Hetch Hetchy there were no gray areas, only black or white.

Both Manson and Muir found an eager outlet for their opposing passions in the newspapers of the day. The San Francisco newspapers, particularly the *Examiner, Call, Bulletin,* and *Chronicle,* vigorously supported the city's interests and the Manson view. When Muir claimed that the use of water and power from Hetch Hetchy would advance "predatory interests," the *Call* called Muir to task, noting that the only predatory interests would be the needs of a million people. Muir's ideas were "the conservation of natural resources run mad. . . . Mr. Muir fails utterly to understand the policy and pushes it to absurdity." The *Bulletin* admitted that most women were "commonly supposed to be on the side of sentiment," but the editor believed that sensible women would not be taken in by Muir's foolish notions about nature.[78]

While it was nearly impossible to read a good word on Muir in the San Francisco papers, this was not the case nationwide. The *New York Times,* the *Boston Transcript,* and various newspapers picked up the Hetch Hetchy story. William Colby and Eastern members of the Society for the Preservation of National Parks saw to it that editors received their press releases, as well as letters to the editor by such men as Muir, McFarland, Johnson, Allen Chamberlain, and Colby. Many of these newspapers responded with pro-preservation editorials, much to the annoyance of the San Francisco editors, who viewed Hetch Hetchy as a regional issue. Such outbursts they considered meddling by the ill informed.

As the year 1909 came to a close, even prominent San Franciscans sought compromise. The Commonwealth Club of San Francisco offered a forum with the hope that some sort of middle ground could be reached. The club membership met fortnightly for lunch, and an invited speaker presented a policy issue of significance. Although dominated by business interests, the club prided itself on a logical, impartial approach to controversial issues. Hetch Hetchy was the subject of more than one meeting. The club president even appointed an investigating committee, a step sometimes taken for issues of special importance. The Hetch Hetchy committee, headed by Beverly Hodghead, assisted by E. A. Walcott, Frank Adams, and J. K. Moffitt, reported to the membership in November 1909. Whereas San Francisco politicians and City Engineer Manson fully expected the committee to rubber-stamp the city's plans, they were astonished at the endorsement of the Sierra Club position. The committee made three recommendations: (1) Lake Eleanor should be fully developed, (2) "Hetch Hetchy should be held as a resource of the bay cities to be used after the above sources have been developed to their full capacity," and (3) all camping privileges and other uses must remain with the people.[79]

Members read Harriet Monroe's aesthetic description of the valley as a "garden of paradise" where the shining river "wandered lazily . . . turning back upon its course, tangling itself into S's and M's as if it were loath to leave so beautiful a place." The report analyzed the Hammond-Hall claims to Lake Eleanor water, recommending that they be resolved. The committee did not address land issues within the valley, except to note that the natural attractions would be permanently lost should the city persist in its plans. The message was clear: WAIT! The city should not be in a hurry. There was plenty of water if the city developed Lake Eleanor and, presumably, Cherry Creek. Hetch Hetchy Valley should be reserved, leaving a future generation to decide its fate.

Of course this solution was not new. Both Roosevelt and Pinchot had advocated such a compromise. Muir and the Society for the Protection of National Parks were more than willing, indeed eager, to grant the city Lake Eleanor. The Commonwealth Club, however, not only prided itself on advocating growth and beautification but also usually supported the side of business. To advance any policy that might hamstring economic progress could border on sacrilege. Whether the report represented the majority view of the membership went unrecorded, but the surfacing of a report that opposed the views of city leaders surely suggested that the residents were not so committed to the Hetch Hetchy project as commonly thought.

The new secretary of the interior, Richard Ballinger, soon realized that Hetch Hetchy represented a hornet's nest of conflicting views. Given the situation and beset with petitions to void the Garfield grant, he took the recommendation of the Senate Committee on Public Lands and others to authorize a study. The study would give him time and ease the pressure, no matter which direction he might turn.

E. G. Hopson and Louis J. Hill, both engineers with the U.S. Geological Survey, conducted the research and compiled the study that reached

FIGURE 8. The *San Francisco Call* often lampooned John Muir and the "nature lovers" as impractical, feminine, and foolish. Above all, the newspaper claimed, the valley defenders had little understanding of technological progress in an advancing civilization. *Courtesy of the Bancroft Library.*

Ballinger's desk in December 1909. The budget to produce the work was totally inadequate. Thus, the two engineers had to rely greatly on information from Manson and his city engineer's office. According to the two men, San Francisco was not forthcoming. Hopson wrote his boss, George O. Smith, that the information provided by the city was "practically nil." He expressed incredulity that a great city could base its decision for a costly system "on the strength of the few scanty facts that have been obtained." His partner, Louis Hill, was equally disenchanted, noting that he was "surprised to learn how little San Francisco really knows about the water supply she desires to acquire or of other possible sources which might be available." Not only were they uninformed, but the city engineer's office seemed to be devious and lacking in frankness.[80]

Like the Commonwealth Club report, the official Hill and Hopson study recommended that Lake Eleanor–Cherry Creek be developed, assuring a storage capacity of 350,000 acre-feet. If the city needed more storage, the Poopenaut Valley, located a mile below the Hetch Hetchy Valley, could be dammed, providing an additional 30,000 acre-feet of water. In concert these sources should mean that the city would not have to use the Hetch Hetchy Valley for at least 50 years. Over the Christmas holiday season and the New Year, Secretary of the Interior Richard Ballinger contemplated what he should do.

CHAPTER 5

San Francisco to "Show Cause"

"By care in the designs, the use for water supply can be made to add greatly to the scenic value."

THE FREEMAN REPORT

In early February 1910 J. Horace McFarland, head of the American Civic Association, wrote an intriguing letter to William Colby. "If you were sitting at my desk," he exclaimed, "I could tell you in full detail some things I cannot write about the Hetch Hetchy situation." He had just left the office of Secretary of the Interior Richard Ballinger, and "if I was William E. Colby, E. T. Parsons, and particularly if I was John Muir," he wrote, "I would try to read through the lines of this letter and between them, and see that there is strong hope that the right will triumph."[1] By the end of the month the secret was out. Ballinger had ordered San Francisco to "show cause" as to why the Hetch Hetchy Valley should not be deleted from the Garfield grant. The city would have to provide conclusive proof that it needed the valley—not an easy task. It was a triumphant moment for those who had worked to safeguard Hetch Hetchy. John Muir wrote Ballinger that "all the right-minded people throughout the country and particularly out here, are rejoicing over the stand you have taken for the preservation of our National Parks."[2] For the valley defenders it was a high point, and as historian Holway Jones noted, "it appeared that the fight was won and that nothing . . . could stem the tide of victory."[3]

At San Francisco's City Hall, the mayor, the city engineer, and various officials were both disappointed and belligerent. They immediately requested a copy of the Geological Survey report by Louis Hill and E. G. Hopson, which formed the basis for Ballinger's decision. They also asked for a hearing and even had the audacity to set the date.[4] The secretary, of course, was not to be bullied about, and while granting the city a hearing, *he* would set the place and the time: Washington, D.C., May 18, 1910.

Once more the momentum had swung, this time to those who wished to shield the valley from San Francisco, reserving it for tourism and future generations. Environmental disputes are not won by the rightness of a cause—if that can be determined—but more often by money and persistence. Defenders of Hetch Hetchy won the first round when Secretary of the Interior Ethan Hitchcock concluded that the sanctity of Yosemite National Park could not be violated. He denied San Francisco a permit. However, rather than retiring, city leaders hesitated momentarily and then moved forward with renewed energy and determination. The tide turned in 1908, when Secretary Garfield granted the city the use of the valley. By this time, however, John Muir and his Sierra Club were fully awakened to the threat. The messianic leader for scenic lands and the national parks led a nationwide campaign to stop the San Francisco steamroller. Now, with Ballinger's "show cause" announcement, Muir's victory seemed close at hand. His forces delivered a knockout punch and all that was necessary was the referee's count to ten. San Francisco, however, returned to its feet by count five and rallied to begin a devastating counterpunching attack. With power, political savvy, commitment, a dynamic mayor, and unlimited amounts of money, the city would take the offensive with a talented new city engineer and a grand Hetch Hetchy water proposal designed by the top civil engineer in the nation. With very limited funds, the Sierra Club and the Society for the Protection of the National Parks did the best they could to defend their position, but the city once again regained the momentum as the fate of the valley moved toward resolution.

Both sides prepared for the May 1910 hearing before Secretary Ballinger. While San Francisco city attorney Percy Long and city engineer Marsden Manson prepared a brief to substantiate the city's case, the defenders of the valley made their own plans. In early March Colby suggested that Muir come down and stay at his Berkeley home so that the two might plan, anticipating San Francisco's arguments. As the two sides readied, the unequal nature of the fight became evident. San Francisco had a permanent staff, a lobbyist, money to pay any consultant who might prove helpful, and a budget that allowed for comfortable travel and first-class lodgings at Washington's Willard Hotel. The Sierra Club had little, save for the sympathies of certain Appalachian Mountain Club members and the support of leaders of the American Civic Association. The club had to appeal for contributions and depend on pro bono legal work and on the willingness of reasonably affluent members and friends to contribute their time and money. Although this group would not label itself radical, in many ways its members were "radical amateurs": middle-class, white, well-educated men and women who volunteered endless hours in a cause they believed in. For most of the men and women it was a moral crusade, and their work, dedication, and occasional success became "the driving force of conservation history."[5]

One example of such dedication was young William Frederic Badè, who held the chair of Old Testament literature and Semitic languages at the Pacific School of Religion in Berkeley. He was an avid member of the club and so devoted to John Muir that following the "prophet's" death, Muir's daughters asked that Badè become the executor of Muir's literary work. Eventually the professor edited five volumes of Muir's writings, while also serving as editor of the *Sierra Club Bulletin*.[6] In 1910 Badè was enjoying a sabbatical year of research in the British Museum and around Europe. In spite of his scholarly endeavors, he kept close tabs on the Hetch Hetchy situation and even lectured on the issue to his fellow ship passengers and anyone else who would listen. When Badè heard of the upcoming hearing, he wrote Colby that he was willing "to cross the pond" to Washington, but he had no money. Colby responded that the hearing was crucial and "we must have you there. In a few days I will send you about $200.00." It was time "to strike a telling blow," and Badè's testimony could be unique and timely since he had consulted with recreational experts in Switzerland and elsewhere.[7] Rather than aesthetic themes, the religious studies professor would defend the valley with solid economic arguments, anticipating its development as a tourist center.[8]

Both sides understood that engineering expertise would be important. Through statistics, San Francisco would have to prove its need, and the Sierra Club would have to prove otherwise. To accomplish its task, Colby needed to find a competent engineer, one able to dispute the facts and figures of City Engineer Manson and his staff. Did San Francisco truly need the water? Did the city need to submerge the valley to fulfill its future water requirements? The eternal subtlety between want and need would be central to the debate.

The club's difficulties in finding an engineer willing to testify is revealing. After consulting a number of candidates, Colby and Joseph N. LeConte settled on Charles Gilman Hyde, a highly respected hydraulic engineer and professor at the University of California.[9] Hyde agreed and asked for reimbursement for travel, food, and lodging and a fee of $15 a day. But then a strange metamorphosis weakened the professor's resolve. On May 6, only 12 days before the scheduled hearing, Hyde wrote Colby that he had worried himself sick and had concluded finally that he should not testify. His friends and colleagues had advised him against involvement, and university administrators were fearful that the school's good name might be brought into the matter. He could not undertake work that "would jeopardize the best interests of the institution." Colby understood what had happened. In a letter to Muir he surmised that "President [Benjamin Ide] Wheeler probably persuaded him that it would be detrimental to the University for him to represent us."[10] *Persuade* may be too gentle a word, for Wheeler was a good friend of Gifford Pinchot and had privately and publicly supported San Francisco's aspirations. Furthermore, he did not believe that professors were free to speak their minds. Academic freedom had limits, and Wheeler insisted that "the university should be harmonized to the demands of its constituency."[11]

With Hyde silenced, Colby searched the fraternity of hydraulic engineers, a tight little knot not easily loosened, especially when asked to testify *against* an engineering project. He wired Edmund A. Whitman, the attorney and leader in the Appalachian Mountain Club, to find him an engineer ready to testify for $500. No luck. He finally found Philip E. Harroun, a Berkeley engineer familiar with Hetch Hetchy. But Harroun asked for $1,050 to testify, and Colby confessed that "the money situation is going to be rather a hard one to handle."[12] Harroun testified and received his salary, but it took more than six months to scrape together the full payment.

After a delay to May 26 to accommodate the schedules of the army engineers, the appeal hearing got under way. Harroun did a commendable job in presenting statistics showing that San Francisco had no real need for Hetch Hetchy water for years to come. Manson, of course, disagreed. He pleaded that the city had not had time to collect all the necessary hydrological data but, clearly, Hetch Hetchy was necessary for its water project to succeed. Edmund Whitman objected. Manson had had plenty of time to gather data. The city just hadn't done it. Whitman and others argued a four-point program that included the heart of the defenders' solutions. The secretary should revoke the Garfield grant but confer Lake Eleanor–Cherry Creek water rights to San Francisco. This watershed land should be removed from Yosemite National Park and placed in the adjacent Stanislaus National Forest, thus leaving park land inviolate. Finally, a code for the entire national park system should be enacted to prevent future encroachment on the parks' dedicated purpose.[13]

The preservationists' program offered a reasonable solution, and point 4 certainly anticipated the National Park Act of 1916. But Ballinger, after hearing all testimony, followed a moderate course by granting San Francisco time to make a study of other water sources sufficient for the Bay Area. Also, the city must design its proposed Hetch Hetchy system and evaluate its damage to the scenic features of Yosemite National Park. In essence, Ballinger instructed San Francisco to prepare what today we would call an environmental impact statement. The city must submit this statement to a special Board of Army Engineers, which would make a recommendation to the interior secretary.

Although the burden of proof still lay with the city, those who hoped to shield the valley from development were disappointed. They had hoped that Ballinger would raise roadblocks, forcing the city leaders to abandon the Hetch Hetchy Valley and search out other available water sources. They had hoped to strike, as Colby put it, "a telling blow," but now the city had escaped and even acquired a new ally—time.[14]

Meanwhile, changes and charges on the national political stage impacted Hetch Hetchy. By the summer of 1909 Gifford Pinchot sincerely believed that Secretary Ballinger was dismantling the Roosevelt-Pinchot conservation policy. He did not hesitate to speak out, particularly with a flurry of charges that Ballinger had made inappropriate grants of coal lands in Alaska. This clash, which became known as the Ballinger-Pinchot controversy, was, in the opinion of Pinchot's biographer, responsible for the loss of authority of both Pinchot and Ballinger, and eventually President William Howard Taft.[15] In September 1909 Pinchot and President Taft met privately in Salt Lake City. Pinchot said he could not change his often hostile opinions and he would not be muzzled. Taft countered that he would not like to lose "Pinchot's valuable service to the nation." Reflecting on that meeting, Pinchot realized that Taft "might be forced to fire me." In a letter dated January 7, 1910, Taft did just that, charging that Pinchot had acted like a muckraker without evidence. The forester, whom historian Elmo Richardson has called "a White Knight—righteous, audacious, and restless," was out of office, but not invisible. He would still be influential in the Hetch Hetchy controversy. Ballinger was somewhat vindicated in a subsequent congressional investigation that filled 13 volumes of testimony. Yet his authority was undercut and his motives became suspect. Some believed that his request that San Francisco "show cause" was nothing more than a political move to enrage Gifford Pinchot—an avid supporter of the city's claims. Discouraged by criticism and failing in health, in March 1911 Ballinger sent Taft a letter of resignation, which the president accepted.[16]

President Taft, perhaps intent on appointing a low-profile secretary of the interior, chose Walter L. Fisher of Chicago. His views on conservation were little known, but on the surface the new secretary would probably favor San

Francisco's interests, for he was good friends with both Garfield and Pinchot, and he was seen as a "commodity conservationist."[17] Yet the valley defenders were not disheartened by Taft's choice. In a reassuring letter to Colby, Stephen Mather, soon to become the first director of the National Park Service, mentioned that he had known Fisher for nearly 20 years and that he felt sure he would be a splendid man for the post. Mather went on to say that he had gathered a few people to talk over the Hetch Hetchy situation and that "Mr. Fisher was one of them and he had a good chance to hear our side of the case." Any decisions from him "will be the outcome of a fair and unbiased judgment."[18] Harriet Monroe agreed.

One of the first letters that Secretary Fisher received was from San Francisco city attorney Percy Long, who wished to inform the secretary that "the City expects to be governed by Mr. Freeman's conclusions and judgment as to the presentation of this case [Hetch Hetchy] in response to the order to show cause."[19] It seemed that significant changes were being made in San Francisco as well as Washington. The city had hired engineer John R. Freeman as a consultant to further the Hetch Hetchy project and essentially replace Marsden Manson. It was a propitious move by the city, one which would eventually checkmate the valley's defenders. For San Francisco a new chapter opened with the appearance of Freeman, whom President Taft had recommended to the city and who was, perhaps, the most prestigious hydraulic engineer in the United States. Based in Providence, Rhode Island, he had assisted in the design of water works in Boston, New York, and other major cities. He had consulted with William Mulholland in the construction of the Los Angeles aqueduct system. He commanded respect among his fellow engineers, and opponents and politicians were wary of questioning his engineering conclusions. Even conservationists had a certain trepidation in disputing his assumptions regarding aesthetics and landscape. If San Francisco was ever to win the right to dam the Hetch Hetchy Valley, Freeman was the professional to do it. He had informally consulted with San Francisco leaders well before January 1912. But now San Francisco would be governed by his judgment, at the princely salary of a retainer of $2,500 a year plus a per diem of $200 a day when away from his home office.[20]

One of the results of Taft's appointment of Walter Fisher and San Francisco's employment of John Freeman was that the whole democratic decision-making process for Hetch Hetchy slowed down. Fisher wanted time to become better acquainted with the issue. Freeman wanted time to design his Hetch Hetchy system and write his report. Furthermore, Marsden Manson had suffered a nervous breakdown in early 1912 and went on a reduced workload. Before his retirement, Ballinger had set another Hetch

Hetchy review for December 1, 1911, but Fisher postponed that date. With the interior secretary amenable to San Francisco's requests for time, the year 1912 became one of granting extensions so that the city might prepare its case. Its advocates were somewhat confused, struggling with a strategy. Perhaps they should consider a direct appeal to President Taft? Some supervisors wished to go directly to Congress.[21] The scheduled March meeting came and went, as Fisher rescheduled the meeting for June and then November. Since the city used Manson's health as the reason for asking for delays, the valley defenders questioned whether Manson's breakdown was genuine. After the June hearing was canceled, J. Horace McFarland wrote that he hoped that "we can finish the 'knockout' in the fall, although if Manson can again be conveniently ill, he may secure more time." McFarland suggested that it would be good to know the extent of Manson's illness and just how much he was absenting himself from his office. Writing from Boston, journalist Allen Chamberlain edged closer to the truth when he suggested that Manson's illness and the subsequent delays look "like an effort to put off action until after the presidential election." Some feared that the delay allowed Representative John Raker of California, an avid proponent of San Francisco's claims, to introduce a bill to grant San Francisco the valley. However, both Fisher and Allen Chamberlain assured the West Coast supporters that no legislation would precede the hearings scheduled for December 1.[22]

While Manson appeared to be incapacitated, John Freeman was not. During the first half of 1912 the talented engineer and his staff had compiled a massive document that fulfilled Ballinger's and the army engineers' request that San Francisco demonstrate the need for water, indicate how it would be obtained from the Tuolumne River watershed, and assess what its possible damage to the scenic features of Yosemite National Park might be. In short, the Freeman Report would "show cause" why San Francisco should have the Hetch Hetchy Valley. It was a massive, 401-page document printed on heavy, lacquered pages, suitable for the reproduction of photographs. The volume weighed five pounds, eight ounces.[23] The thick, red-covered volume proved a turning point for the city.

What the report articulated most brilliantly was the artful suggestion that a reservoir in the Hetch Hetchy Valley, far from damaging the beauty of Yosemite Park, would actually enhance it. It would become a destination place for pleasure-bound tourists. The first 52 pages displayed the beauty of still water, particularly reservoirs. With a skillful narrative and large-scale photography, pages depicted reservoirs in Norway, the British Isles, and the eastern United States. As one turned the pages, it was evident that utility could be blended with beauty, valleys improved by dams, and reservoirs enhanced by roads. Eastern reservoirs, such as a brimful Croton Lake, featured walkways with couples dressed for a turn-of-the-century afternoon stroll. If a visitor preferred a less strenuous activity, "hundreds of automobiles tour around this artificial lake each pleasant Sunday afternoon in summer." For further proof of the salutary effects of reservoirs, Freeman offered Lake Thirlmere, the water storage unit for Manchester, England. At one time opposed by locals, the roads around the reservoir "have come to be one of the most popular holiday routes in England, and are much traveled by coaching parties, automobiles and groups of cyclers."[24]

In his photography Freeman showcased a touched-up Hetch Hetchy reservoir, treating the reader to the scene of a totally calm lake reflecting the cliffs and waterfalls in the background. Another virtual photograph showed roads on both sides of the reservoir with the caption, "View of the lake from near the proposed hotel site." Replacing the word *reservoir* with *lake,* the report promised that the outlet pipes in the dam would be placed "so high that the lake never could be wholly emptied," thus partially avoiding the "bathtub ring" associated with reservoirs at low water.[25]

At the same time, Freeman depicted the valley in its present condition as a very miserable place. According to U.S. Army figures (the army administrated the park up to 1916), only 269 visitors had frequented Hetch Hetchy since 1909, many of them San Francisco engineers and employees, and Sierra Club members who enjoyed summer outings. Freeman suggested that this dearth of visitors was, in part, a class issue. Since no roads existed, a couple would have to reserve a week of time, then hire horses and a guide, thus spending around $200 to visit the valley. Few Americans could afford such a vacation. Once there, one either was too hot or was plagued with mosquitoes. In other words, not only was the unimproved Hetch Hetchy Valley expensive to visit, but it was not a pleasant place for humans and never would be unless transformed. Once the city built a road, the great valley would be physically and economically accessible to average Americans intent on touring the Yosemite and the Hetch Hetchy Valley. Freeman asserted that once an attractive reservoir had cured the mosquito problem, the length of the tourist season would triple.[26]

The report maintained that Hetch Hetchy was merely a poor substitute for Yosemite Valley, admittedly beautiful but relatively tame and uninteresting compared to its grand and varied counterpart. Why not create something new and original? "The flooding of the valley floor . . . would present features different from anything found in the Yosemite or elsewhere in California." Furthermore, the waterfalls would remain. In essence, the Hetch Hetchy Valley's value was in its scenery and as a splendid water supply. "By

FIGURE 9. This springtime photograph of the valley shows in the right foreground the small lakes that Whitney found charming. Engineer John Freeman and dam advocates saw them as mosquito factories that made the valley uninhabitable. *Courtesy of the Yosemite National Park Archives and Library.*

care in the designs, the use for water supply can be made to add greatly to the scenic value," creating a beautiful lake in a spectacular valley that thousands of people would enjoy in the next century.[27] Hetch Hetchy as a reservoir, in Freeman's estimation, could serve both the needs of the "nature lover" community and that of San Francisco for an adequate supply of pure water.

Part 2 described the new design of the whole Hetch Hetchy system, including the dam, development of water storage at Lake Eleanor and Cherry Creek, a downstream electrical power station, an elaborate underground pipe system, and the 150-mile aqueduct featuring an ingenious gravity system that eliminated the need for pumping stations. The Freeman plan was more grandiose than the Grunsky plan (1902) or the Manson plan (1908–1911), but Freeman argued that "in ten years' time the state of the art of water supply has advanced and the San Francisco problem has broadened."[28] Furthermore he envisioned a population explosion that would reach 3,632,000 for San Francisco, Oakland, and the surrounding Bay Area towns by the year 2000.[29] While the Freeman system would initially create a water surplus, in time the anticipated 400 million gallons per day would be needed.

Freeman was also expected to evaluate all other possible water sources. In other words, he had to examine the many other rivers that flowed down the long Sierra Nevada slope, as well as a couple of lakes and the Sacramento River. In an impressive section of 21 appendices, he found all of them inferior to that of the Tuolumne River in quantity or quality of water, in hydropower production, or in the ease of acquiring water rights. Contesting the Spring Valley Water Company data, the report questioned whether the private company could fulfill the future needs of the city.[30] Later some of this New England engineer's data would be compromised by revelation of his shoddy research methods, but for the lay reader it was most convincing, particularly since Freeman was, as historian Donald Jackson notes, "never reluctant to remind businessmen that they were ill-equipped to comprehend the complexity of technological systems."[31]

The appearance of the Freeman Report left the defenders of Hetch Hetchy in a momentary state of shock. Robert Underwood Johnson wrote Colby that "I have the big book . . . and am appalled by the enormity of the publication." McFarland called it "a most portentous document." Harriet Monroe had received "San Francisco's enormously weighty report" and confessed that even if she could "read every word of the voluminous report, I probably should not know what to say in reply." William Colby, so industrious and committed, would soon write a brief in response, but even he was overwhelmed by the heavy red book, confessing in mid-October that he had read some of Freeman's work but only scanned much of it.[32] In sheer size the work was daunting. No one knows exactly how much money the city of San Francisco spent in salary and production costs, but the report was worth every dollar. Kevin Starr was right when he noted that "most reports are filed and forgotten. Some have a partial effect on the decision-making process. A few reports, the very few, envision and help materialize the future."[33] The Freeman Report was of the latter variety.

It fell to William Colby to respond. In a lengthy statement, Colby argued that commercial interests should be refused entry into the national parks. He attacked the objectivity of engineer John Freeman, submitting that he simply wanted to enhance his client's cause. Freeman was an engineer for hire, and all his statistics reflected his obligation to the city of San Francisco.

The Colby Brief effectively argued that the future value of the Hetch Hetchy Valley lay in tourism, not water storage. In 50 or 100 years "the need of the Nation for Hetch Hetchy Valley and the extensive camp and hotel sites on its floor will be greater than the need of San Francisco for its use as a reservoir site."[34] Of course both sides played the tourism card. The Freeman Report favored campgrounds and at least one hotel situated above a pristine, reflecting reservoir with roads skirting each bank. The Colby Brief's plan, however, was more extensive. It envisioned a developed valley, complete with hotels, restaurants, campground, trails, horses, and fishing opportunities, all within the lush valley and the ever present cliffs and waterfalls. Freeman suggested limited human use, while Colby favored extensive use of the immediate region. To emphasize the potential commodity value, the Colby Brief looked to Switzerland, where the Alps were "practically unfrequented at the commencement of the 19th Century and now its visitors are counted by millions."[35]

Particularly surprising was the Colby Brief's advocacy of the valley's potential for winter sports. In a pre–downhill skiing era, winter sports consisted of ice skating, tobogganing and sledding, and a crude form of cross-country skiing—none designed to attract thousands of participants. Yet the report stressed that with a good road and proper facilities, recreationists would "turn to Hetch Hetchy *both in winter and in summer*" (emphasis in original). It suggested that Secretary Fisher and Congress again consider comparison with Switzerland, where year-round resorts "were crowded with tourists and the hotels were filled." The number of winter visitors to Yosemite would increase by 1930 to between 30,000 and 50,000 people "if facilities of access are provided." How would these new recreationists be accommodated? "Hetch Hetchy is the only other valley in the Park that ranks with Yosemite in point of availability for hotel sites; it is the only one comparable to

Yosemite for the sublimity of its scenery." Some might question the significance of winter sports, but by 1916 at Huntington Lake, tourists enjoyed a winter carnival featuring skating, snowshoeing, and skiing; horse-drawn sleds; and, of course, blazing fires.[36] If there was any question regarding the valley's potential tourist value, Colby erased doubts by stating that "Switzerland's enormous profits from tourist travel show that an economic factor is involved here of such proportion that it will ultimately knock San Francisco's soap and power profits into a cocked hat."[37]

The key to this development was roads. As earlier mentioned, in 1907 the Sierra Club officially advocated a road into the valley and continued to champion a highway from the west.[38] Surprisingly, the Colby Brief went further by endorsing a road from the Hetch Hetchy Valley climbing southeast to Tuolumne Meadows, thus bisecting the great canyon of the Tuolumne River. The brief argued that a reservoir would block that possibility and would "obstruct the natural highway through the Tuolumne Canyon to the Tuolumne Meadows."

Today, such a suggestion would be greeted as sheer heresy not only by environmentalists but by the general public. But the defenders of Hetch Hetchy wanted to open the region to all people, and they were aware that few Americans would visit by foot or horseback. In many ways, the turn-of-the-century "wilderness cult" was more sentiment than reality. Most people who read Muir's mountain adventures considered the vicarious experience sufficient. If they were to visit mountain country, they wanted roads to get there. The Sierra Club was quite willing that they should have them. Even John Muir saw a place for them. *The Mountains of California* opens with the almost classic observation that "thousands of tired, nerve-shaken, over-civilized people are beginning to find out that going to the mountains is going home; that wildness is a necessity." How would they get there? Muir said by "means of good roads."[39]

In spite of Colby's effort to neutralize the Freeman Report, San Francisco took the offensive. If there was any doubt of the fact, a midday meeting on August 31, 1912—at the same time as the release of the Freeman Report—between Mayor Jim "Sunny" Rolph and civil engineer Michael M. O'Shaughnessy would confirm it. The place was the Whitcomb Hotel on Market Street, the temporary city hall. Rolph had just been elected, and he was enjoying his popularity. He was a friend of business interests, having served twice as a trustee of the Chamber of Commerce, and also a friend to labor, having supported collective bargaining during the waterfront strike of 1906. However, his most successful accomplishment came with a humanitarian effort after the earthquake and fire, when he opened his home to the

Mission Relief Association, feeding thousands of homeless people.[40] Resplendent in his pin-striped pants, cutaway, and cravat and with a boutonniere in his lapel, Rolph was determined to hire a new city engineer.[41] Stretching his legs and feet, encased in his ever present cowboy boots, the mayor spoke of the challenges still present in reconstructing the earthquake-ridden city, but his primary concern was the water system. O'Shaughnessy listened intently. The Hetch Hetchy system, especially Freeman's plans, appealed to the ambitious civil engineer, a man who wanted to place his mark on the world through a great engineering project.[42]

However, O'Shaughnessy was not an easy sell. According to his own recollection, San Francisco had not treated him well. He had come from Ireland, having graduated with honors in engineering from the Royal College of Dublin, to San Francisco, working first as a reporter on economic issues for the *San Francisco Chronicle.* However, he was soon working at his engineering profession. He spent time at the Big Island of Hawaii, building a rather remarkable canal, known as the Olokele irrigation ditch, for a private sugar cane estate.[43] Soon he returned to work in San Francisco, but he felt that the city had "cheated" him out of his $5,000 fee after he provided surveys to extend Market Street to the Pacific Ocean. The following year he had a similar experience with the extension of Potrero Avenue, in which his fee "went a-glimmering." Furthermore, he knew the difficulties of politics in the city. In 1908 he had granted a *Los Angeles Times* reporter an interview in which he characterized San Francisco leaders as "featherweight political demagogues."[44] At that time he was engineering dams in the San Diego area, and he would stay there. When Mayor Rolph wired him for an interview, however, he traveled north, doing so without even demanding what he believed the city owed him.

What attracted O'Shaughnessy back to San Francisco? Certainly working for Mayor Rolph had its appeal, and he did have a strong personal reason. His wife, Mary Spottiswood, wished to return to the city of her childhood. Furthermore, not all his Bay Area consulting had turned sour. He enjoyed working with Francis Newlands in the design of Burlingame, as well as the engineering challenges he faced in planning and constructing San Francisco's California Midwinter International Exposition in 1893–1894. Furthermore, a humanitarian reason may have influenced his decision. San Francisco was still reeling from the earthquake and fire, and his talents could make a difference. But it was the engineering rigor of the Hetch Hetchy system that cinched the deal. Perhaps San Francisco occasionally had wounded his pride, but pride could be overridden by ambition. Hetch Hetchy would challenge his skills until his death in 1934. The project would be his chance to lead in the construction of one of the great civil engineering challenges of his day, one that would open the whole Bay Area to development. He was inspired by not only civil engineering but also *social* engineering. Mayor Rolph concluded the deal with a pep talk: "Go to it, it's up to you, you must look on the City as your best girl and treat her well."[45]

There seemed to be no time to spare. After lunch Rolph ushered his new city engineer to the St. Francis Hotel to meet John Freeman. Perhaps the two engineers had met professionally, but at this hotel meeting, the two began an enduring friendship. Certainly part of the friendship was professional admiration. O'Shaughnessy believed the Freeman Report for Hetch Hetchy was close to engineering perfection. In the years to follow he enjoyed consulting with Freeman, tweaking the plan here and there, but basically carrying out the engineering challenges. Before engineering, however, came politics. With Rolfe in attendance, the two discussed the forthcoming hearing with Secretary of the Interior Walter Fisher. They also examined the Garfield grant and the needs of the city. Without question the upcoming national election also gained their attention. November would be a crucial month.

In the meantime Michael O'Shaughnessy wanted to visit the mountains. He was a field engineer, not one to spend time only at his desk. It was important to see, and examine, the Hetch Hetchy Valley through the vision of his engineering eye. By railroad and wagon he traveled to the Hog Ranch, nine miles from the valley. On September 16 the new city engineer mounted his horse for the ride with his guide, Hank Williams. On the trail they encountered a large party coming out, headed by Taft's secretary of war, Henry Stimson. The secretary wanted to discuss Hetch Hetchy, suggesting the main valley could be saved by a dam in the upper valley. O'Shaughnessy tempered his response but later rejected the idea as "very undesirable." Once he entered the deep canyon and flat-bottomed meadow, two city employees, Charles Hill and E. J. Koppitz, met him. Both were in their thirties and doing water measurements while bunking in the city-owned cabin, one of four structures in the valley. For the next two days O'Shaughnessy walked the valley, noting the vertical walls of granite and exploring the bedrock on each side of what would become the dam site. The meandering, deep-flowing river drew his attention, and no doubt he made mental notes on how it would be diverted if he built the dam. He was excited. One friendly account of his visit indicated that "his eyes sparkled. . . . The engineer triumphed. He saw a perfect spot, a lake-site made by nature. One dam in that narrow pass, one small dam not three hundred feet wide at its base, would hold back in that granite bowl the flood waters of the river, would pour them out again for the city's need, and the lands of the San Joaquin. In that gigantic natural reservoir could be stored water for four million people, water for four hundred thousand acres of fertile farms."[46] Clearly this natural receptacle could be made to serve the needs of man. On September 18 O'Shaughnessy left the valley to visit Lake Eleanor and Cherry Creek, jotting notes on the potential for a low dam and on the properties and water rights that the city had purchased from William Hammond Hall. He returned to San Francisco on October 19 and, the next day, held a press conference in which he endorsed the Hetch Hetchy dam site and plan.[47]

Ironically, only one day after O'Shaughnessy left the valley, a larger, more illustrious party entered. Led by Major James Forsyth, the group included Secretary of the Interior Walter Fisher; Robert Marshall of the U.S. Geological Survey; Marsden Manson, the recently retired city engineer (who had regained his health); and J. Horace McFarland, representing the valley's defenders. Fisher considered the three-day pack trip a pleasant informational event. He brought along Robert Marshall as a personal adviser, while Manson and McFarland would present opposite points of view. It was a new venture for J. Horace McFarland, whose experience with nature was a matter of horticulture, specifically tending his rose gardens. However, as he later wrote Colby, camping in Yosemite and Hetch Hetchy "seems to agree with me, for I certainly never felt better in my life than at this moment."[48]

However, not all was perfect. Although Fisher had just left a successful conference in Yellowstone and was certainly primed for the beauties of nature, what he first encountered at Hetch Hetchy were difficulties and discomfort. Colonel Forsyth turned what should have been a leisurely horseback trip into a forced march, resulting in a secretary of the interior who, McFarland noted, was "completely prostrated with the rough and unnecessarily dusty and dirty trip." By evening he had recovered, and for an hour around the campfire the party debated the ultimate fate of the valley that encircled them. Fisher listened, but commented little. Marshall and McFarland judged the discussion "a draw," not bad considering the difficult travel day. In fact, an exhausted McFarland recorded that he "would have given Hetch Hetchy Valley to anybody for a two-cent postage stamp that night."[49]

Early the next morning Colonel Forsyth insisted, with what McFarland described as "the usual perversity which I find characterizes army officers," that the party ride the steep trail to examine Lake Eleanor. The lake, beautiful enough, would require an 1,800-foot-wide dam to impound a small amount of water. Manson admitted that the city wouldn't develop Lake Eleanor unless it could get Hetch Hetchy as well. Manson talked nonstop, and McFarland and Marshall let him do so, assuming that his continual prat-

FIGURE 10. The perfect V of the river outlet suggests why civil engineers found the valley such an attractive reservoir site. To them, it seemed created for such a purpose. *Courtesy of the SFPUC.*

tle would do his cause more harm than good. In fact, McFarland quipped that if Manson's words "could be transmuted into water, there would never need be any doubt as to the capacity of San Francisco's supply." Returning to the valley, some took a swim in the river, but evidently tired of endless talk, Fisher struck off on a solitary, lengthy hike up the valley. He took a dip in the river, viewed the waterfalls and the magnificent meadows and cliffs. The valley spoke to him in a less contentious way, and he returned refreshed. At the evening campfire no one raised the contentious issue, and in McFarland's memory, "all were of one mind as to the exquisite beauty of the place and as to the undesirability of ever giving it up."[50]

The last day, they rode out of the valley and then boarded the rough Yosemite stage. On the ride out, Robert Marshall, who was a member of the

Sierra Club and opposed to the dam plan, set a trap into which Manson innocently jumped. Marshall suggested that Manson should ask him, in the presence of Fisher, if the city might obtain the Poopenaut Valley (a smaller valley just below the Hetch Hetchy through which the Tuolumne River flowed) to dam and further increase storage capacity for San Francisco. Manson did so, but with poor results. Angrily, Secretary Fisher turned and snapped, "What *other* valleys in the Yosemite, Mr. Manson, is it the idea of the city of San Francisco to absorb?"[51] For once, silence prevailed.

McFarland felt that Fisher had been captivated by the magnificence of the Hetch Hetchy country and that Manson had not made a strong case for San Francisco. Marshall, while forced on the trip to feign neutrality, wrote Colby a note that he had "the pleasure of flatly contradicting Mr. Manson on several occasions."[52] No one could read the tea leaves regarding Fisher's attitude, but like innumerable visitors to Hetch Hetchy, he had been enthralled by the beauty, isolation, and solitude of the place. One should remember, as well, that while Manson represented San Francisco, he was really a minister without a portfolio. Whatever impressions he might have made on Fisher, he had been replaced by Michael O'Shaughnessy, and the interior secretary knew it.

As the Interior Department hearing approached, two other interested parties produced and submitted massive documents. The cities of Oakland and Berkeley employed engineer J. D. Dockweiler to report on the East Bay water sources. The Dockweiler Report, more than 500 pages long and partially funded by the city of San Francisco, emphasized the solidarity of the East Bay and San Francisco with regard to their future water needs. At this point San Francisco was courting the two cities with the knowledge that East Bay water needs would enhance the city's ability to "show cause."[53]

William Bourn, head of the Spring Valley Water Company, ordered the other report. The city still depended on Spring Valley water, and Bourn wished to make it clear that under his leadership the company had become respectable and had grown with the demand, expanding to capture a large watershed in the hills of Alameda County, to the southeast and across the Bay from San Francisco. Now Bourn wished to show Secretary Fisher, as well as the city supervisors, that Spring Valley could provide for the present and future needs of San Francisco and its suburbs. To do that, he hired the well-known former army engineer Hiram M. Chittenden, who in turn employed a dozen engineers to produce an equally long report of more than 500 pages, proving that Spring Valley Water Company was quite capable of supplying the needs of San Francisco for many years to come.[54]

Thus by November 1912 the Interior Department had three massive reports to peruse. It was a battle of the bulk. However, the Freeman Report clearly had the edge. It was not that the research was better or the facts more accurate. It came down to a matter of presentation and promotion. Although the Hetch Hetchy issue was not yet before Congress, the city of San Francisco distributed the book to every representative and senator. In all likelihood the vast majority never read the detailed engineering material. What they, or perhaps their legislative assistants, did view were the many impressive photographs in the first 50 pages. As the congressional hearings would later reveal, many congressmen accepted Freeman's argument that the valley's beauty could be enhanced by a reservoir.

Secretary Fisher finally convened the hearing on Monday, November 25, 1912 beginning six days of testimony and argumentation, some of it heated. Accounts from both sides understandably differ on what took place. O'Shaughnessy recalled that Fisher vigorously cross-examined some of the "nature lovers" to uncover their state of mind. The city engineer, effusive in praising the testimony by representatives of the city, poured a good deal of scorn on the opposition. He aimed his acidity particularly at the testimony of Robert Underwood Johnson, whom he delighted in calling "Underbrush Johnson." Johnson did not seem to have a good solution to San Francisco's need for water. When Fisher quizzed him about what the women and children of San Francisco would do, he responded that the city would have to "condense the water of the Pacific Ocean." O'Shaughnessy was amused, knowing that such an idea was prohibitively expensive. Such an impractical position annoyed Fisher, and later Horace McFarland noted that "our good friend Robert Underwood Johnson took an absolutely indefensible position and stuck to it, thereby losing all influence with the Secretary."[55]

Alden Sampson, an avid mountain climber and member of the Sierra Club, made an effort to keep William Colby informed through telegrams. On November 25 he wired that City Attorney Long had "Presented Case Well Enough," but that Mayor Rolph had "Made a Donkey of Himself." He did not think that the city had gained much ground in the first two days. Two days later McFarland wired that Freeman had "Seriously Discredited Himself by Evasions" and that a member of the army engineers "Had Brought out Failure of Freeman to Supply Information—Whole Situation Encouraging." On the same day, Badè offered his opinion that it was "Hard to Tell Outcome—May Grant Hetch Hetchy with Severe Restrictions—Spring Valley and Irrigationists Still to Be Heard—Reported City Will Try Congress."[56]

The last two days of testimony featured heated exchanges between the attorneys from San Francisco and the Modesto and Turlock irrigation districts.

In dispute, of course, were questions regarding the volume of water that flowed in the Tuolumne River, the irrigation districts' legal right to that water, and the guarantees that San Francisco would be required to make before the districts would sign off on the Hetch Hetchy plan. In their distrust of San Francisco they often found themselves strange bedfellows with the nature lovers. Alden Sampson said he was "betting on our side." J. Horace McFarland, however, was not so sure. He wrote Mather that from what he heard at the conclusion of the hearings, "the Secretary rather abruptly put it up to San Francisco to get busy," since the Board of Army Engineers required more information before it could make a recommendation.[57]

At the conclusion of the hearing, Secretary Fisher turned over all of the testimony and supplementary reports and information to his advisory board, employed by the Army Corps of Engineers. This board would make recommendations to the secretary and his staff. Fisher would rely heavily on their opinion, for he judged their views the only objective, impartial ones. Both sides understood the importance of the army engineers' report, and both hoped that its conclusions would vindicate their position.

Although Fisher considered the board objective and free from bias, all three were military men and all three were engineers. Colonel John Biddle headed the board, with the assistance of Lieutenant Colonel Harry Taylor and Major Spencer Cosby, but H. H. Wadsworth, as assistant engineer in the U.S. Army, carried out the actual investigation. One might assume that the U.S. Army might have leaned toward the city's position, but that would not be true. Ever since 1896, when Lieutenant Colonel S. B. M. Young had arrested a San Francisco party for carrying loaded rifles in defiance of park rules, the relationship between the city and the military had remained strained. Superintendents colonels Harry Benson and James W. Forsyth had kept in close contact with John Muir and the defenders. Muir acknowledged the army's diligence in protecting Yosemite, declaring "blessings on Uncle Sam's soldiers! They have done their job well, and every pine tree is waving its arms for joy."[58]

Significant for the outcome was the board members' calling as professional engineers. They belonged to a fraternity that commanded great respect and authority. Engineers were the technical experts who supposedly knew best how to deal with questions of natural resources. They made decisions by the force of their calculation and their facts, which few could dispute. Colby and the valley's defenders were quite willing to defer to their expertise.[59] No professional group paraded a longer list of qualifications. When the *San Francisco Examiner* printed the résumé for John Freeman, it exceeded the column inches allotted to many of the paper's other stories.[60]

Questioning the objectivity of an engineering board was rare indeed. However, when, in 1909, the Senate Committee on the Public Lands proposed an impartial board of engineers to investigate the Hetch Hetchy situation, both sides embraced the idea. The only dissenting voice came from Robert Underwood Johnson. To Senator Newlands's question about whether a report by the Army Corps of Engineers would be a good thing, Johnson responded: "I should think that would be very admirable, if in addition to that there were included lay members who were impartial, not living in California, but men of such high standing as to render their appointment a guaranty of impartiality."[61] Newlands took note of the suggestion, but the idea went nowhere. Johnson, of course, understood that the army engineers excelled at solving technical problems, but their orientation limited them. A board of engineers was unprepared to answer the concerns of the nature lovers, such as: What of the growing importance of national parks? What of the future use of the Hetch Hetchy Valley as a tourist attraction? What was the direction of American culture regarding leisure time, outdoor recreation, and a need to escape the burgeoning cities? Army engineers were ill-equipped to address these social and cultural questions.

One other factor mitigated in favor of the city. Money was not a problem for San Francisco, but it was for the Interior Department. When Secretary Richard Ballinger originally ordered the report by the Special Board of Army Engineers, he allocated no funds for such a purpose. Instead, he quietly requested that the city pay for the study and furnish information to the board on demand. This financial arrangement could give the appearance of guaranteeing a result in San Francisco's favor.[62] Neither Ballinger nor Secretary Fisher, however, seemed concerned with such a potential conflict of interest.

On February 13, 1913, officials released the long-awaited report. The conclusions represented a victory for San Francisco, although the Hetch Hetchy Valley defenders did note that the engineers provided other water options. The board accepted Freeman's conclusion that in the year 2000 the population of the Bay Area communities would be 3,632,000 people, requiring 540 million gallons per day (MGD). Given that the present Spring Valley water facilities provided 133 MGD, the new source should provide about 400 MGD. The Hetch Hetchy system would be adequate and would be approximately $20 million cheaper than any other feasible project that could furnish the 400 MGD. Most of the other rivers and sources failed the test to deliver the needed water. Furthermore, the report suggested that "the necessity of preserving all available water in the Valley of California [Central Valley] will sooner or later make the demand for the use of Hetch Hetchy as a reservoir practically irresistible." Given this conclusion, there seemed no reason to delay construction. However, the report also concluded that other options existed. The engineers noted that "the best available sources, outside of the Tuolumne River, appear to be the filtered Sacramento [River], the McCloud [River], a combination of the Lake Eleanor–Cherry [Creek]–Stanislaus [River] and Mokelumne [River], and a combination of the American-Consumnes-Stanislaus, and Mokelumne [Rivers]."[63]

The engineers ignored the national park status of Hetch Hetchy and gave no real weight to scenic value. Given the criteria they established, their conclusion to allow construction was self-evident. Colby, Muir, and other defenders had an unwarranted confidence that the engineers could give a balanced opinion. However, when the board treated Hetch Hetchy Valley land as just another site rather than one located in a national park, its conclusion surprised no one. Perhaps the defenders realized their error when, in later years, both Colonel John Biddle and Lieutenant Colonel Harry Taylor kept in close contact with Michael O'Shaughnessy, each visiting the dam site on more than one occasion to heap praise on the city engineer and the work crews.[64] A major reason why the Sierra Club and its allies ultimately lost the fight was that they failed to influence the selection of the board and to shape the questions that the "impartial" board asked. The board, consequently, discounted cultural and environmental issues. Scenic value, tourist use, and national park invasion—the issues on which the Sierra Club based its case—all fell outside the Board of Army Engineers' purview. And, it goes without saying, neither the Sierra Club nor the Society for the Protection of National Parks could afford to produce its own report.[65]

In spite of the advantage of San Francisco, Secretary Fisher hesitated to move forward with a dam decision. Finally, three days before leaving office, he sent Mayor Rolph a letter that surely disappointed him. Fisher determined that he lacked the statutory authority to grant a permit to San Francisco. He decided, in other words, that he could not make a decision. Therefore, the order to the city to "show cause" would remain in effect. He and his predecessor interior secretaries had overstepped their authority in the matter. Recourse for San Francisco to change the order must come from the Congress of the United States.[66]

The chilled champagne of celebration would have to wait for another day.

CHAPTER 6

Congress Decides

"The Destruction of the charming groves and gardens, the finest in all California, goes to my heart."

JOHN MUIR

THE DEMOCRATIC PROCESS can be a fickle one. Elections bring changes—some unplanned—that determine outcomes, including the fate of a relatively obscure valley. The interplay between the executive and legislative branches is cumbersome, seemingly dedicated to slowing down the process of decision making. At times both sides in the Hetch Hetchy controversy might have wished for a benevolent dictator to decide in their favor. Interior Secretary Walter Fisher could have acted in that capacity, but instead he deferred to Congress, perhaps knowing that the whole issue would end up there anyway. It did.

By 1913 the time had come for a decision. Both sides had perfected and presented their arguments. San Francisco based its hopes for water and power needs on John Freeman's idea that the city's victory could provide water and power, while giving the American people a scenic lake set deep in the granite frames of the Hetch Hetchy Valley. The city believed that through good planning and design the practical needs of a growing city could be linked to the aesthetic standards of a national park. The Sierra Club and the Society for the Preservation of National Parks did not deny the city's right to a bountiful water supply. However, they vigorously argued that there were other viable options. The Hetch Hetchy Valley should remain undefiled—revered as scenery, protected by the mantle of the national parks, and maintained as a place where thousands of Americans could refresh themselves while spending their dollars.

Of course, as 1913 unfolded, there was much speculation about who President-elect Woodrow Wilson would select as his secretary of the interior. Defenders of the valley preferred Secretary Walter Fisher, but they knew that "the new President will want to make a clean sweep."[1] He did indeed. Wilson choose Franklin K. Lane for the Interior post, a San Franciscan well acquainted with Hetch Hetchy. He had served as city attorney between 1900 and 1903, and during his tenure he worked closely with Mayor Phelan to obtain water rights in the valley. It was Lane who appealed to Secretary of the Interior Ethan Hitchcock to reverse the secretary's decision to deny the city a permit. If there was any doubt of Lane's position, he resolved it within a week of taking office. The new secretary called in San Francisco city attorney Percy Long and an attorney representing the Modesto and Turlock irrigation districts, with the intention of hammering out their differences regarding Tuolumne River irrigation waters. Richard Watrous, McFarland's assistant at the American Civic Association, heard of the meeting and asked that he be included. Lane extended an invitation, although he would have preferred otherwise. At the meeting, Watrous recalled, the new secretary stated flatly that as city attorney he had favored the Hetch Hetchy scheme, he had favored it in the intervening years, and now that Wilson had elevated him to secretary of the interior, he would continue the battle in and out of Congress. Watrous could only compliment the secretary for his honesty, knowing that Lane's zeal would be very hard to overcome.[2]

The Lane appointment came as a surprise to all parties. The preservationists knew little about President-elect Wilson. In their view this Eastern-born, Princeton University president had come to power when both Taft and Roosevelt floundered. Woodrow Wilson had no record on conservation matters, and presumably little interest. If he had an opinion regarding the philosophical split between Gifford Pinchot and John Muir, it was to favor the utilitarian camp. Presumably Wilson appointed Franklin Lane as interior secretary because the San Franciscan worked hard on his campaign and delivered both the city and state to the Democrats. Much later Horace Albright, second director of the National Park Service, would dispute that interpretation, convinced that "Franklin Lane's appointment to the cabinet was made specifically for the purpose of pushing this [Hetch Hetchy project], the so-called Raker-Pittman Bill."[3]

Albright's recollection is questionable, for there is no evidence that Wilson had ever met Franklin Lane or had even heard of the Hetch Hetchy controversy. Colonel Edward House, a wealthy Texas cotton farmer, had become the close confidant of Wilson. Known for his political acumen, House would be the one to search for a new secretary of the interior. His diary gives us his thoughts and some clues regarding the appointment. On November 21, 1912, House spent most of the day with Lane. They discussed the position, but Lane indicated that he was happy with his present situation and would not change it. Their conversations turned to James Phelan, whom Colonel House described as "a man worth from four to ten million dollars, a bachelor, a fine after-dinner speaker, able but cold blooded." Evidently the perceived latter trait and the fact that he "has some enemies" were enough to eliminate Phelan. House did note that although the West wanted Phelan, the East would not. Although House never mentioned Hetch Hetchy, he did indicate in a memo to Wilson that "the East is all for conservation and the far West is for it in a limited way—that is where it does not conflict with their material interests."[4]

Colonel House continued his search. He considered Walter Page, Wilson's lifelong friend, the editor of *World's Work,* and the man soon to be appointed ambassador to Great Britain. But Page was a Southerner being considered for a position traditionally occupied by a Westerner. In the meantime, Lane had second thoughts. He asked House if he would have a free hand in running the department. Colonel House assured him that if Lane proved capable, Wilson would have no cause to interfere. In the middle of February, House recommended Lane for either "War or Interior." Wilson quizzed his adviser on Lane's views on conservation. House didn't know but said he would invite Lane to Washington, and during the meeting Norman Hapgood, the muckraking editor of *Collier's Weekly,* could "drop in and let them meet as if by accident." House incorrectly believed that the outspoken editor was a conservationist who represented "the extreme Eastern view." Lane, presumably, passed the Hapgood test. Wilson and House continued to jockey their cabinet posts, but finally on February 24, just a little over a week before the inaugural, House offered Franklin Lane the Interior post, and he accepted it.[5]

From Colonel House's diary we may surmise that the ongoing Hetch Hetchy fight had little to do with Lane's appointment. It is possible that Colonel House and/or Norman Hapgood may have mentioned the conflict, but it would have made little difference, since Hapgood supported the city's position. Whatever the exact nature of the San Franciscan's appointment, he was the antithesis of an impartial interior secretary. Lane had an agenda of water development in and out of the national parks, and he would follow it. One cannot help but note the serendipitous nature of his ascendancy, seemingly so unimportant to the new Democratic administration but so crucial to a small glacial valley tucked away in the Sierra Nevada mountains.

For San Francisco Lane could not have been a more perfect choice. Yet, in a sense, the city forces were frustrated by Secretary Fisher's "midnight decision" that he did not have the authority to rule on a permit and that San Francisco must go before Congress. The Fisher decision angered Michael O'Shaughnessy, who, after reading a copy, marched into the outgoing secretary's office and harangued him, saying all the copies of his decision belonged in his waste basket.[6] If Fisher had not tossed the ball to Congress, Secretary Lane could have quickly reversed Ballinger's "show cause" decision, restoring the Garfield grant. By the summer of 1913 O'Shaughnessy could have been busy clearing the valley, beginning construction, and enjoying a cool river swim in the evening. But because of Walter Fisher's resolution to throw the whole issue to Congress, O'Shaughnessy remained in Washington.

Thus, to the city engineer's displeasure, the "nature lovers" would be given one more chance. But they would be on the defensive. San Francisco city attorney Long and city clerk John S. Dunnigan soon drafted a bill authorizing San Francisco's use of the Hetch Hetchy Valley. They prevailed on Congressman John E. Raker of California to introduce it and enlisted the influential support of Congressman William Kent. Fortunately for San Francisco, Wilson felt it necessary to call Congress to a special summer session to consider tariff reform. Under the spurious argument that the city suffered from a water shortage, Raker managed to get the Hetch Hetchy bill on the calendar. Muir sent a telegram to Henry Fairfield Osborn that the "San Francisco Schemers Making Desperate Efforts to Rush Bill Through Congress. . . . You Know the President and Am Sure You Will Strike Hard." Letters resulted from Muir's appeal, but the bill appeared quickly before the House Committee on Public Lands, and for weeks San Francisco's lobbyists had been preparing the committee.[7]

O'Shaughnessy, Long, and Dunnigan were only part of a phalanx of San Franciscans who descended on Washington and took up residence at the Willard Hotel. James Phelan, Rudolph Spreckels, and Alexander Vogelsang—the last soon to be appointed by Lane as assistant secretary of the interior—were often seen at the capitol. John Dunnigan led the effort to draft what would become the Raker bill. They all stayed at the Willard Hotel, suffering through the humid heat. In the evening, to escape his stuffy room, O'Shaughnessy frequently walked, ending his stroll at the ice cream parlor. Often he was accompanied by Congressman James M. Graham from Illinois, who happened to be on the House Public Lands Committee, a person whom he "became sincerely attached to." Presumably O'Shaughnessy used the word "sincerely" to avoid the reader's suspicion that he was lobbying, but friend or not, Congressman Graham would play a significant pro–San Francisco role in the hearings.[8]

Certainly most of the members of the House committee were predisposed to give San Francisco what it desired. With skillful lobbying and Secretary Lane's administrative assistance, the city lined up the support it needed. The San Francisco delegation testified for the bill, buttressed by the appearance of Secretary Lane, but strong support came from Chief Forester Henry Graves, Gifford Pinchot, and directors of the Geological Survey and the Reclamation Service. Pinchot, always influential, pronounced his dictum that "the fundamental principle of the whole conservation policy is use," and he could see no "reasonable argument against the use of this water supply by the city of San Francisco." With regard to aesthetics, Reclamation director Francis H. Newell proclaimed that "there is nothing more beautiful than a well built dam with a reservoir behind it." James Phelan would second Newell's assumption, arguing that by "constructing a dam at this very narrow gorge . . . we create not a reservoir but a lake, because Mr. [John] Freeman has shown that by planting trees or vines over the dam, the appearance of the dam is entirely lost." The concept of beauty offered by Newell and Phelan would be repeated again and again as San Francisco's congressional supporters excerpted the Freeman Report to prepare their statements.[9]

Edmund Whitman, speaking for the Eastern Branch of the Society for the Preservation of National Parks, gave impassioned testimony, but he had never visited the Hetch Hetchy Valley and was clearly outmaneuvered by the assemblage of pro-city supporters. The defenders of the valley were particularly injured when William Denman, a San Francisco judge and charter member of the club, testified that the Sierra Club was quite divided with regard to Hetch Hetchy, negating any impression that the club spoke with one voice. The anti-dam forces did not arrive in Washington. They had been caught off guard by the suddenness of the hearing, and political communication seemed to be like many of the defenders—on vacation. Raker's ability to get the issue quickly before the House in the middle of summer was a strategic bit of political finesse that practically guaranteed success.

The only chance of slowing the city juggernaut came in early July, when Eugene J. Sullivan, president of the Sierra Blue Lakes Water and Power Company, testified before the House Public Lands Committee. Sullivan argued that the Mokelumne River could fulfill San Francisco's water needs and that former city engineer Marsden Manson had suppressed a report that would prove his contention. The committee members were interested, but later, when asked to produce the report, Sullivan lamely explained that it was not yet ready and his assistant, engineer Taggert Aston, was ill and unable to come to Washington. On direct questioning, Sullivan reluctantly named city engineers C. E. Grunsky and Manson as two of the conspirators who buried the report.[10]

During the lunch break Sullivan spoke with his lawyer, and by the afternoon his accusations had softened. Congressman James Graham of Illinois, O'Shaughnessy's ice cream parlor pal, was also busy consulting with the city engineer. At the appropriate moment, Graham attacked, plunging into Sullivan's background and successfully revealing him "as a petty confidence man whose word was utterly worthless." However, two crucial statements in Sullivan's testimony turned out to be true. It was unfortunate that his past was so vulnerable. The report on the Mokelumne River was real, researched and written by a Max J. Bartell, a young assistant in the city engineer's office. Manson did suppress it, but on the basis that it would be difficult to secure the necessary water rights. Not until the 1920s, when the cities of Oakland and Berkeley would tap the Mokelumne River for their water supply, was the truth of the study revealed.[11] In retrospect, the Sierra Blue Lakes Water and Power Company presentation was a viable alternative. But again, the messenger was every bit as important as the message. In 1906 the city had turned away the Bay Cities Water Company proposal because it was tainted by the graft trials. The Blue Lakes proposal died because Eugene Sullivan made a fool of himself and, in turn, made the proposal look foolish.

Support for Hetch Hetchy defenders came from Representative Halvor Steenerson of Minnesota. He attacked Freeman's presumptions regarding reservoirs, stating that during low water visitors would view "a dirty, muddy pond . . . and probably some dead fish and frogs in it."[12] Steenerson also raised the issue of a dominant, imperial San Francisco holding sway over the hinterlands. He was "opposed to the eternal drawing upon the Federal Government resources and of the people to make cities more attractive at the expense of the country." He admired such men as Muir, Robert Underwood Johnson, and Liberty Hyde Bailey because they "are deploring the fact of the influx to the city from the country." Steenerson defended sentiment, giving the legislators an aphorism to think about: "It is a wise saying that the man who writes the songs of a people has more influence than he who writes its laws."[13] With his "back to the land" philosophy, Halvor Steenerson voted against the wishes of the city.

All of this debate seemed to matter very little. The House Public Lands Committee members had congealed to the point that they were almost hostile to the nature lovers. No valley defenders could make the trip, and the Eastern advocates were suspect as they called for preservation of an unfamiliar area. Many congressmen agreed with Representative Charles Thomson of Illinois who confessed "not to have a great deal of patience with gentlemen who sit by their firesides in the East and discuss the plan . . . without a better knowledge . . . of the facts." The Public Lands Committee moved the bill onto the House. On September 3 the Raker bill won approval by the full House by the comfortable margin of 183 to 43.[14]

The San Francisco delegation were pleased. They assumed that President Wilson would sign the bill into law, which left only the Senate to frustrate their plans. Michael O'Shaughnessy and Percy Long went home after the hearing, hoping they would not have to come back again. Each, according to O'Shaughnessy, returned to cool San Francisco "lighter in weight by about ten pounds after the torrid Washington temperatures."[15] City Clerk John Dunnigan, however, stayed until the end of the year, working on the Raker bill when necessary and lobbying during the intervening time.

As Dunnigan lobbied and labored on the Raker bill, he found that his most cordial and receptive ally was Congressman William Kent. The Marin County representative opened his home to Dunnigan, inviting him to stay as long as he was in Washington. Dunnigan gratefully accepted. No doubt he found his quarters quite satisfactory, but from the viewpoint of the Hetch Hetchy controversy, it was perfect. Kent had strong preservation credentials, particularly because he had taken the lead in a winning fight to preserve Lake Tahoe from a disastrous private power plan that would lower the lake level, at times, some 50 feet.

In the eyes of Muir and the Sierra Club defenders, Kent's reputation rose to new heights when he purchased a large grove of redwood trees in Marin County and then transferred title to the United States. This spectacular grove, soon to be known as Muir Woods National Monument, was destined to be cut by loggers and then the valley dammed for a reservoir.[16] Kent could not tolerate that possibility. In January 1908 he explained to Muir that the purchase was an "uncontrollable impulse":

> The hideous heedless wickedness of trying to butcher those trees put me in a frame of mind where I wondered, how far a trustee ought to go to protect such a trust. I am sure the danger is passed now and hope I can forgive Jos [James] Newlands and William McGee, who for a few dirty dollars would have deprived millions of their birthright.

Such passionate, moralistic purpose bonded Kent to the older mountaineer. Muir wrote Kent that his magnificent gift would make him "immortal like your Sequoias, & all the best people of the world will call you blessed."[17]

When in 1910 Kent's Marin County constituents elected him to Congress, Muir and the Sierra Club felt certain that they could count on his support. However, William Kent had another side that the valley defenders did not fully understand. When he campaigned for Congress, he assured his neighbors that he did not believe in locking up natural resources. He was a Pinchot conservationist; indeed, he admired the forester and often corresponded with him. One of the principal reasons he ran for Congress was that he opposed the policies of Interior Secretary Richard Ballinger and William Howard Taft. He felt a "giveaway" was taking place and that the concepts of the public lands and public power were being undermined by the current administration. He favored national parks, and even fought for park recognition for Mount Tamalpais, adjacent to his home country. However, tourist value motivated his thought, rather than preservation of a unique ecosystem. He was, in effect, a human-centered conservationist, who admired greatly the Pinchot Roosevelt program.[18]

Not only did his brand of conservation fall on the utilitarian side, but he had compassion for the down-and-out of human society, developed earlier when he lived in Chicago. In the windy city, waste, poverty, and human misery influenced his social philosophy. He became a benefactor of Hull House, and as a Progressive patrician, his sensitivity toward the plight of the poor marked him as an active social reformer. He cared about nature and parks, but he could not turn his back on the needs of San Francisco and its people. Knowing his background, it should not have surprised the Hetch Hetchy Valley defenders to see him defending the rights of the people of San Francisco. Speaking in favor of the Raker bill, he stated: "My idea of conservation would teach that if Niagara Falls could be totally used up in alleviating the burdens of the overworked sweatshops of New York City I should be glad to sacrifice that scenic wonder for the welfare of humankind."[19] Translated to Hetch Hetchy Valley, that sentiment meant that he was quite willing to give the area over to San Francisco, if it would improve the lives of the many thousands who lived there. His concern for urban needs and his commitment to a mildly socialist philosophy trumped his preservation tendencies. Kent's dilemma, of course, represented that of many legislators enamored with nature and the parks but beholden to their constituencies and the public welfare.

Kent's Washington home became the de facto headquarters for the San Francisco organizational effort. Dunnigan was there almost every evening. Phelan, Long, John Raker, and other San Franciscans often joined the two men. They talked strategy. The group not only worked on fine-tuning the Raker bill but also made valuable connections with Kent's political friends, especially among both Democrats and the Rooseveltian group of Republicans. Later John Freeman, who languished in Europe during most of the fight, would credit Kent and Dunnigan for the bill's passage.[20]

After the decisive defeat in the House, the Society for the Preservation of the National Parks, under whose masthead the Sierra Club and Eastern supporters worked, knew they must do better if there was any chance to forestall a San Francisco victory in the Senate. Muir, home from his 1912 world travels to South America and Africa, was again willing to take up the fight, but declining health diminished his energy and his willingness to travel. He did, in May 1913, make the short journey from Martinez to Berkeley to receive an honorary doctorate. As University of California president Benjamin Ide Wheeler heaped praise on the naturalist as a man "uniquely gifted to interpret unto others [nature's] mind and ways," it must have troubled Muir to be hooded and honored by a university president who had strongly favored the city's position on Hetch Hetchy, even to the point of suppressing open debate and the free expression of ideas within his own faculty.[21]

Increasingly the leadership of the friends of Hetch Hetchy fell to Robert Underwood Johnson. Muir encouraged his friend and urged him to "strike hard and very fast." The result was a number of booklets and leaflets which suggested that much more was at stake than just the valley. Although money was tight, the society printed more than 20,000 pamphlets, booklets, leaflets, and circulars. Volunteers mailed them not only to various friends and organizations but also to the entries in *Who's Who in America*.[22] The city of San Francisco, their argument went, claimed almost half the land within Yosemite National Park in its wish to control the entire Tuolumne River watershed. If the city was successful, Yosemite would be effectively cut in half, with hiking and overnight backpacking severely limited.

Besides booklets and leaflets, Robert Underwood Johnson circulated "An Open Letter to the American People." The letter articulated not only the value of Hetch Hetchy but the importance of nature in people's lives. Johnson's most effective argument was to show the folly of San Francisco's argument that an alternative water source would be $20 million more costly:

> Put up at auction, what would this wonderland bring? "What am I bid," the auctioneer might say, "for one superb valley, twenty miles of unique cascades, half-a-dozen snow peaks, beautiful upland meadows, noble forests, etc., now owned by a gentleman named Uncle Sam, suspected of not being able to administer his own property? Do I hear $20,000,000 to start the bidding? Remember, that these natural features are priceless.[23]

Johnson's letter prompted a number of responses. Former president William Howard Taft, then living in New Haven, Connecticut, wrote Johnson a long letter, opening with "I am with you in the Hetch Hetchy matter" and then setting out a strategy.[24] Senator Reed Smoot sympathized with Johnson, but believed that the Raker Act would pass the Senate, mainly because "everyone opposed to it is branded as a tool of the electric light power trust." Just how influential was the so-called electric light power trust? No one has provided a satisfactory answer, although certainly such companies as Pacific Gas and Electric Company opposed the city's public utilities aspirations. Not all responded positively to Johnson. Harvard historian Albert Bushnell Hart seemed to relish tweaking Johnson's appeal, ending his letter: "After all, nature, like the Sabbath, was made for man."[25]

As debate opened in the Senate, the success of the circulars and Johnson's letter was quite remarkable. Senator Reed Smoot informed his colleagues that he had received at least 5,000 letters against the proposal from all over the nation. In support of Smoot, another senator stated that he had received "innumerable letters on the subject." Senator Henry Ashurst of Arizona confessed disbelief that he had received between 3,500 and 4,000 letters on the issue. Senator James Reed of Missouri wondered how a mere two square miles of land could cause such a stir, in which "the Senate goes into profound debate, the country is thrown into a condition of hysteria, and one would imagine that chaos and old night [darkness] were about to descend upon the land." The letters also descended on Secretary of the Interior Lane's office—by the thousands. Young Assistant Secretary Horace Albright and three other secretaries had to reply. They found it necessary to forge Lane's signature, and then explain why Lane felt the grant should be made. Later Albright said, "I hated this job, for I was in sympathy with the protests."[26]

Why was there such an outburst over a valley that, as Senator Reed noted, no one knew? Of course the leaflets and booklets contributed, but that was not the whole answer. Knowledge of an issue and action on that issue are different. Few of us *act* on our knowledge. Yet in the case of Hetch Hetchy, thousands of people did, even though it might have been only a brief note. Perhaps their activism was in response to the violation of a national park. However, few people granted any sanctity to national parks. Most Americans had no idea of the difference between a national forest reserve and a national park. And for good reason, for in 1913 the differences were subtle and undefined. It is more likely that the appeals for preservation, and particularly Muir's and Johnson's eloquent expressions, struck a cord with many Americans. It was not a matter of feeling for parks or wilderness per se, but rather about nature: a nature associated with gardens, with parks, with rivers and valleys, and with natural sounds. In a way, the letter writers were reactionary, wanting to save an environment that was disappearing in American cities.

FIGURE 11. Wapama and Tueeulala falls, situated close together, were wonderfully exciting in the springtime. Here we can see firewood already cut, indicating that the days of the larger trees were numbered. *Courtesy of the San Francisco Public Utilities Commission (SFPUC).*

The writers of letters to save Hetch Hetchy Valley were moved to reassert the importance of nature in their lives. Whether each writer could make a difference was, perhaps, not so important as clarifying their own individual values in a world of industrialization and rapid change. Put together, the thousands of letters represented a growing political and cultural force that would assert itself in the new century.

The response of the American public clearly astonished the senators, but most were not moved to reconsider their pro–San Francisco position. Congressmen John Raker and William Kent, city lobbyist John Dunnigan, and other city representatives had sown seeds throughout the year that they could now reap. By early December Michael O'Shaughnessy, James Phelan, Mayor James Rolph, and others had returned to Washington for the final showdown. They stood ready to assist sympathetic senators in any way they could. When junior senator Key Pittman was scheduled to speak on the issue—his first major debate—he called on O'Shaughnessy. The city engineer coached Pittman, "preparing maps illustrating the irrigation benefits of the dam" that worked to make an effective presentation.[27]

THE FINAL SENATE debate commenced on December 1 and would end on December 6, 1913. The debaters first focused on the question of irrigation, specifically the rights of the farmers who secured their irrigation waters from the Tuolumne River. Senator John D. Works of Southern California carried the standard for the farmers for a day and a half, hammering away at the city and refusing to believe that San Francisco would honor its pledge to guarantee the irrigation districts' water rights. He also claimed to speak for the taxpayers of San Francisco against the "enormous burden of debt" being placed upon them.[28] The Raker bill was legally suspect and morally wrong, he argued. San Francisco should look elsewhere for its water.[29]

Works found support from Senator William E. Borah of Idaho. Born on a Midwestern farm, Borah believed that the project would "destroy the San Joaquin Valley" as a great agricultural breadbasket. The grant would give San Francisco a monopoly of water and power. Studying the reports, Borah was convinced that the Sacramento, the Eel, or the McCloud River could do the job for the city. Perhaps the cost would be $10 million or $20 million more, but Borah believed the additional cost should not be the deciding factor. It made sense for the city to tap a more northerly river source, for the Sacramento River drainage had much more water than the San Joaquin, into which the Tuolumne River flowed.[30]

Senators Borah and Works found surprising support from Senator Asle J. Gronna of North Dakota. A farmer himself, he was sympathetic to the worried irrigators, but he was one of a handful who argued the Society for the Preservation of National Parks' position. He had heard sarcastic remarks about the "nature fakers," but he disagreed, claiming that he "respected and honored those who unselfishly try to protect the property of the people of the United States." He appreciated the scenic value of Yosemite National Park, and he believed one ought not "to commercialize every bit of land and all of the landscapes of our country." He announced that "it is a mistake to destroy the handiwork of God's creation, for that cannot be duplicated." Eloquent as Gronna's speech may have been, it fell on deaf, or more accurately, absent, ears. Delivered in the late evening of December 4, few senators were in attendance.[31]

By December 5 the mood had swung to the San Francisco position.

Senator Marcus Smith, representing the newly admitted state of Arizona, had practiced law briefly in San Francisco. In sympathy with his former city, he lashed out against those opposed to the Raker bill, calling them "financed by venal, selfish interests . . . forgetting or ignoring the wants and necessities of living men and women." He made certain the Senate understood his priorities. "We all love the sound of whispering winds amid the trees, but the wail of a hungry baby will make us forget it for the while as we try to minister to its wants." "You lovers of nature," he declared, "will do well to give less attention to nature's beauties and more sympathy to the wants of men. Love nature all you please, but do not forget its crowning glory—man."[32] Echoing Kent's social philosophy, Smith declared that he "loved trees, but I love men more, far more. To give to a million people a necessary of life and in so doing violate no right of any other person is a duty so evident, and action so imperative, that my earnest support is gladly given to this bill."[33]

Led by California senator George Perkins, the proponents soothed the irrigators and countered the environmentalists. Perkins presented a long list of prominent Californians who supported the bill. Perkins hammered away at the Spring Valley Water Company's hold on San Francisco, capping his argument with a *Los Angeles Examiner* editorial titled "Set San Francisco Free from Monopoly." A private corporation had long held up and sandbagged a great municipality. It was time to get rid of the "robber barons of a benighted age." However, in the give-and-take of the debate, those opposed to the Raker bill heard from William Bourn, the president of the Spring Valley Water Company, the corporation in question. Bourn wired Perkins that the company "has no lobbyist or representative in Washington; is not in communication directly or indirectly with anyone there, and no one in Washington has any authority or warrant . . . to oppose the bill in its [Spring Valley's] interests." Obviously annoyed by the charges against his company, Bourn requested that his telegram be read to the Senate, and Perkins obliged.

The debate revealed somewhat of a sectional split. For instance, Charles W. Eliot, president of Harvard University, wrote in defense of national parks and against the presumptions of San Francisco. However, David Starr Jordan, president of Stanford University and a charter member of the Sierra Club, supported the city. Benjamin Ide Wheeler, president of the University of California, was adamant in his advocacy of San Francisco's position. It would seem that the East wished to preserve and the West to develop. The sectional difference inspired Senator James Reed to quip that "college professors who never have been near enough the Yosemite Park to know anything whatever about it are enlightening us with reference to our duty. The degree of opposition increases," noted Reed, "in direct proportion with the distance the objector lives from the ground to be taken. When we get as far east as New England the opposition has become a frenzy."[34]

A young freshman senator from Nebraska posed very different concerns. In time George Norris would become the great champion of publicly owned electrical power, an issue, according to his biographer, that highlighted his senatorial career.[35] Norris believed that if the Senate eliminated the power part of the Raker bill, there would not be "one-thousandth part" of the opposition to passage. He believed that the majority of the more than 5,000 letters he received were not true nature lovers, but part of an effort by private water and power companies to frustrate the effort of San Francisco. This "great campaign which somebody has carried on that must have cost somebody a great deal of money" has been waged "because San Francisco would now offer competition to private water and power companies." However, the evidence against the electric power companies was not exactly a smoking gun. In presenting detailed data, which must have caused the assembled senators to nod off a bit, the Nebraska senator tried to show that the ties between the Pacific Gas and Electric Company, Southern California Edison Company, and Great Western Power Company constituted a monopoly. If broken, utility rates could be halved. In a moving peroration, Norris called on his colleagues to

> hold up your calloused hands, you senatorial strap hangers, who for years have been riding on cars here and paying for something you did not get. Hold up your hands, and if you can not give relief to the citizens of the District of Columbia from an intolerable situation, at least let your brethren out on the coast get some relief. Defeat this bill and you will receive the plaudits, the acclaim, and the praise of every hydroelectric corporation in the State of California. Pass it and you give into the hands of the people a power that God intended should do some good for man.[36]

Norris, of course, strongly supported Representative William Kent's insistence on section 6 of the Raker bill, which prohibited the city of San Francisco from selling Hetch Hetchy power to any private utility company.

The debate carried on into December 6. By late evening there were no minds or votes to be changed. Everyone had settled on his position or had, in disinterest, simply turned away from the whole controversy. In truth, the evening sessions had difficulty keeping a quorum on the floor. As the hours slipped away, senators attempted to amend the bill or at least delay a vote. Some felt, as did Senator Clarence Clark of Wyoming, that the bill was simply "a log rolling proposition." He introduced an amendment. Defeated.

Close to the end of the day, senators began to shout "Vote, Vote." At 12:00 P.M. they did: 43 yeas, 25 nays, 27 absentees.[37]

The San Francisco delegates, who had been listening intently from the balcony for six days, were jubilant. Victory was finally at hand. After shaking hands all around, they retired to the Willard Hotel to, as O'Shaughnessy put it, "wet our whistles." But by the time they got there, Sunday had arrived, and no liquor could be sold. Still pleased, yet a bit disappointed, they toasted their triumph with cold water—not so satisfying but perhaps more symbolic. O'Shaughnessy wired his wife: "Victory at Midnight. San Francisco Knows How."[38] Within a few days they were riding the rails west, in plenty of time for Christmas. O'Shaughnessy thought of the great project that lay ahead.

The defenders maintained a forlorn hope that President Woodrow Wilson would veto the Raker Act, but the likelihood that Wilson would go against his secretary of the interior on an issue in which he had little investment was slim indeed. On December 19, 1913, Wilson signed the act, acknowledging that "good and well meaning people opposed the act" but their arguments "were not well founded." He had weighed the evidence and was following the public interest.[39]

For those who had fought to spare the exquisite valley, it was a bitter loss, but one that was expected. Their forces had been depleted. McFarland had suffered a nervous breakdown. Muir was not in good health. Colby seemed to tire. There was, as usual, very little money. The Western Branch of the Society for the Preservation of National Parks left most of the work of leaflet circulation, as well as any monitoring of Congress, to the Easterners.[40] Rumors persisted that, consciously or unconsciously, the naive "nature lovers" had been the pawns of the water and power companies. When Senator Norris suggested that "somebody" was financing the nature lovers' campaign, every senator understood that "somebody" to be the Spring Valley Water Company in cahoots with the Pacific Gas and Electric Company. The fact that such assumptions were false mattered very little.

The preservationists were at a disadvantage that could not be overcome. San Francisco's strategy of a well-funded, well-organized effort, especially in 1913, provides an example of what could be accomplished through determination and adroit management in moving a bill through the halls of Congress.[41] From Mayor Phelan through Mayor Rolph, from City Engineer Grunsky through engineer John Freeman and City Engineer Michael O'Shaughnessy, the city spoke with one mind. Outside of a two-year period, from 1904 to 1906, the supervisors, most of the newspapers, and the political leaders kept their collective eye on the prize. They had also made an investment of close to $2 million in water rights land purchases, which they were unwilling to forfeit. The investment in time, energy, and perhaps even ego made the city determined, even to the point of irrationality, to turn a blind eye to alternative possibilities. There were points along the road of this 13-year struggle when the city might have abandoned the Hetch Hetchy plan, but no leader seemed to consider seriously any side roads, at least not publicly.

Sheer luck had something to do with it. No one would consider the San Francisco earthquake and fire as *luck,* but from the point of view of the dam project, it was. Federal officials, politicians, and the nation would find it too difficult to deny the prostrate city its water, especially since some proponents blamed a lack of it for the devastating fire. The appointment of Franklin Lanc, a man who had, with Phelan, initiated the Hetch Hetchy idea in 1901, as secretary of the interior seemed almost serendipitous, as if the desires of San Francisco had unconsciously influenced the president-elect's choice. Lane's appointment seemed to suggest a cyclical view of history, with the Hetch Hetchy issue coming full circle from 1902 to a dramatic conclusion, one that pleased the proponents committed to "the greatest good for the greatest number."[42]

For the valley defenders it was a hard defeat. However, they tried to view the loss as a skirmish in which they learned winning strategies to be used in the century to come. They came out of the fight bruised, but much wiser politically. Richard Watrous lamented that the "end has come, at least on this particular fight," but he believed they had fought well and were much more sophisticated in the ways of protecting nature and the national parks.[43] For those who had actually spent summer and fall days in the valley, it was an emotional blow. Muir wrote Vernon Kellogg, his Stanford zoologist friend, that "the loss of the Sierra Park Valley [is] hard to bear. The destruction of the charming groves and gardens, the finest in all California, goes to my heart."[44] Muir's intimate affection and attachment to the *place* made the image of bulldozers stripping the valley of all life a devastating thought. He could accept that the valley would change with roads, campground, many people, and hotels, but the metamorphosis from river to reservoir, of stripping the oaks, the pines, and meadows for still water, depressed him. It was fortunate that he never lived to see that day.

Some writers have suggested that the loss of Hetch Hetchy killed Muir. Robert Marshall visited him in Martinez and wrote that it was sorrowful to see him in his "cobwebbed study in his lonely house . . . with the full force of his defeat upon him." Marshall wished that Congress and the president had waited on any action "until the old man had gone away—and I fear that is going to be very soon."[45] He was suffering from depression over his lost valley, but just as much from the weather. Influenza, or the "grippe" as he called it, was his annual winter malady. On January 12 he wrote his daughter Helen that he was enduring "black foggy rainy lung-choking weather," lamenting that "this day is one of the darkest of the melancholy winter." He had been "grippe-choked [for] several weeks."[46] He did find the will to carry on. He took some solace that "some sort of compensation must surely come out of this dark damn-dam-damnation." To another friend he wrote: "I'm glad the fight for the Tuolumne Yosemite is finished. . . . Am now writing on Alaska. A fine change from faithless politics to crystal ice and snow."[47] He knew that time would vindicate the rightness of their cause. And he understood the enduring nature of fundamental issues. "The battle for conservation will go on endlessly. It is part," he wrote, "of the universal warfare between right and wrong."[48]

Perhaps as devastating as the loss of the valley was the passing of loved ones and friends. Louie Strentzel Muir, his wife, had died in 1907. William Keith, Scotsman, artist, fellow mountaineer, and friend, passed away in 1911. John Swett, so helpful with Muir's frustrations with writing, left this world in 1913. Edward Parsons, so loyal to the Hetch Hetchy fight, died in early 1914. His friends were leaving him, and he increasingly relied on his daughters, Wanda and Helen, for support, both emotional and physical. But they could do nothing regarding their father's failing health and lung infection. Muir finally succumbed on Christmas Eve, 1914, just over a year after Wilson signed the Raker Act.

In the meantime City Engineer Michael O'Shaughnessy studied Freeman's grand Hetch Hetchy plan. If all went well on the political and financial fronts, the determined Irish engineer would soon transform the valley into its final use.

CHAPTER 7

To Build a Dam

"I give waters in the wilderness and rivers
in the desert to give drink to my people."

ISAIAH 43:20

THERE IS AN excitement to building, to actually creating what you have dreamed and planned on paper for years. Michael O'Shaughnessy felt that intensity when he returned to San Francisco. He knew that the Hetch Hetchy Valley and the immense water system would occupy his thoughts for the next decade. As a dirt-under-the-fingernails engineer, O'Shaughnessy never felt comfortable as a lobbyist in Washington. At one point, thoroughly tired of the food at the Willard Hotel, City Attorney Long offered to treat the delegation to a famous local restaurant known for its delicacies, including frog legs. When the waiter nodded to him, O'Shaughnessy ordered plain ham and eggs, which, he recalled, "caused considerable mirth and consternation amongst my less democratic associates." But the Irish civil engineer shook off any criticism, remembering that he longed "for the smell of the construction camp."[1]

With the Raker Act in place, he would now be able to get back to construction camps. The 49-year-old engineer, soon known simply as "the Chief," would get his chance to build a great dam and a complex water system, one that would fulfill his professional desire to make a difference and also partially carry out James Phelan's wish to make San Francisco and the Bay Area the premier region of the American West. As construction began in 1915 and finally concluded in 1934, there was no end of controversies. It is impossible to detail all of them. The scale of John Freeman's plans would test even a crusader's zeal. It is an unwise property owner who builds a 5,000-square-foot home when he needs only 2,000. Freeman's Hetch Hetchy plan was far too grandiose for San Francisco in 1920. Considering that the Spring Valley Water Company provided every gallon of water for San Francisco and the peninsula until 1934, it is obvious that when the Hetch Hetchy system came on line, it practically flooded the city. Freeman, for all his engineering skill, missed a design opportunity. In retrospect, he might have given the city a more functional system if he had projected to 1950, with following increments to be funded through the growth of the city and particularly the surrounding communities. In his defense, however, he believed in a regional water system, one that would serve the East Bay. But Richmond, Berkeley, and Oakland did not join in, leaving San Francisco with a costly project producing a commodity that was not needed. Electric power production may have saved the city from financial disaster and possibly bankruptcy, for it did have value. The following narrative celebrates the construction of a great water system, but one that could have been planned and performed differently, saving San Francisco a good deal of anguish.

O'Shaughnessy first picked a planning and engineering staff from a host of applicants.[2] He also did his best to heal the wounds of a decade of controversy. J. Horace McFarland gracefully admitted defeat, writing O'Shaughnessy that he felt it unnecessary "to add any to the general water supply by shedding large amounts of post mortem tears."[3] Once hired, his staff began the job of securing land and rights-of-way for reservoirs, aqueducts, power development, and a railroad. Easements across private lands had to be obtained and/or purchased for the 167-mile aqueduct. Water rights, always a contentious subject, had to be clarified. Such tasks kept City Attorney Robert Searles running to the law library to keep covetous men at bay. Landowners were always anxious to profit at the city's expense. Searles spent much time in court defending the city against suits such as that brought by the owner of a mining claim along the railroad grade near Moccasin Creek who demanded $100,000 for the right-of-way. The presiding Tuolumne County judge awarded the recalcitrant landowner $160.[4]

Related to the immense size and capacity of the system, was the difficulty of financing the project. Although the voters of San Francisco had generously passed a $45 million bond issue in 1909, the city could not sell the securities. Six years later $43,394,000 in bonds were still unsold.[5] Construction could not begin until the actual cash was deposited with the city treasurer. Funding the Hetch Hetchy project would remain a more difficult hurdle to clear than its actual construction. Speaking to a local group in 1915, O'Shaughnessy predicted completion of the Hetch Hetchy system by 1919, if he had the funding and was free from political change.[6] Of course such a foretelling was wishful thinking, and O'Shaughnessy knew it. Politics and finance would interfere. Still reeling from the earthquake and fire, San Francisco always seemed plagued by an empty purse.

Besides dribbling funding, construction at Hetch Hetchy faced inflation. With the onset of World War I in 1914, prices began to rise. With the entrance of the United States into that war in 1917, both wages and general prices reflected an accelerating inflation. O'Shaughnessy had figured his labor costs at $3 per worker per day, but by the time he was ready to employ workmen, the going rate had risen to $4.50. And even at those wages, men were not always available. The wartime draft drained off his laborers, although O'Shaughnessy was able to wrangle a deferment for some of his workers on the basis that the electric power component of the system contributed to the war effort. The same inflationary spiral applied to the supplies and equipment required for construction. For many items prices rose 75 percent between 1913 and the end the war.[7]

But inflation or not, many believed the Hetch Hetchy project was far too big and far too expensive. It was "a colossal job," "a stupendous folly" with many "blunders and extravagances" that had wasted the $45 million bond issue of 1909 and required constant spoon-feedings of additional money. In 1919 Eugene Schmitz, the disgraced mayor of San Francisco, questioned the wisdom of the project. Schmitz, who seemed to have more political lives than a cat, had been accused of fraud in 1906, along with Abraham Ruef, the notorious city boss. His charisma resurrected his political career, and by 1919 the city supervisor felt free to criticize the size and the cost of what was going on in the Sierra Nevada mountains. He was particularly annoyed that O'Shaughnessy had, at one point, estimated the cost of the Hetch Hetchy dam at about $3.5 million, whereas, in fact, the city paid Utah Construction almost double that figure. With the planned addition, the cost would be at least $10 million. While the tremendous outlay burdened the city, Schmitz estimated that it would be 10 to 15 years before San Francisco saw a drop of Hetch Hetchy water.[8] The project constantly required new funding, to the point that it was popularly joked that O'Shaughnessy's initials, "M. M.," meant "more money." Beyond such amusement, the Hetch Hetchy project constantly faced the danger of a shutdown, not because of labor strife, bad weather, or engineering problems, but for lack of funds.

Yet those problems were largely ahead of them as O'Shaughnessy's staff of engineers and planners tackled the many difficulties and uncertainties of building a dam deep in the mountains. Before even thinking of building the Hetch Hetchy dam, the city had to complete a transportation infrastructure.

FIGURE 12. Before the railroad, teamsters and loggers began work in the valley, cutting and transporting sawlogs to the mill. The lumber was used in San Francisco's many buildings necessary for the undertaking. *Courtesy of the SFPUC.*

City attorneys secured the necessary rights-of-way and engineers laid out the route of a nine-mile road into the Hetch Hetchy Valley, giving access to automobiles and trucks. City engineers surveyed a 58-mile railroad, snaking its way through the Sierra Nevada foothills. They saw to the orderly transformation of tiny Groveland into a thriving construction town. At the Hetch Hetchy dam site, San Francisco planned and built a camp for at least 400 workers. Soon city workers drilled a 1,000-foot tunnel to divert the Tuolumne River, and then built a cofferdam to protect the construction site from flooding. All of these preliminary projects, and many others, had to be completed before the city could accept bids by dam contractors.

The first displacement of dirt, rocks, and trees came when the Utah Construction Company took on the task of building the nine-mile road into the valley, replacing what had been a trail. Although the U.S. Army believed the road would cost $52,000, the city accepted a bid at $180,000. Why so much?

FIGURE 13. Financing the Hetch Hetchy project proved a difficult task. Here Mayor James Rolph (seated front row, center) and City Engineer Michael O'Shaughnessy (to Rolph's right) pose with a number of supervisors and Bank of Italy executives. The commitment of A. P. Giannini and the Bank of Italy, which evolved into the Bank of America, assured that the project would move forward. *Courtesy of the SFPUC.*

The roadbed was 22 feet wide, graded, and surfaced so that one side of it would soon be the railroad bed. When it was completed, trucks immediately began to bring in construction materials and equipment. By September 1915 city carpenters began nailing together the Hetch Hetchy camp, designed to feed and sleep hundreds of workers. The first building was a dining room large enough to seat 500 men, followed by a substantial bunkhouse to replace the tents that gave temporary summer shelter. In quick succession the carpenter crew erected a cement warehouse of 14,000 square feet, a three-room cottage, a hospital, wood house, oil house, meat house, and various other buildings. A water tower and two-inch line, plus a road system, completed the camp, home to many workers for almost six years. Downstream

FIGURE 14. A group of city supervisors and engineers return from Lake Eleanor. O'Shaughnessy always said the supervisors were much braver and cantankerous in city hall than on this road to Lake Eleanor. *Courtesy of the SFPUC.*

from the camp, telephone workers strung lines and loggers built a sawmill at Canyon Ranch, both essential for the coming dam construction. O'Shaughnessy reported that his men took great care to preserve the natural forest appearance at the ranch, as by leaving a screen of trees immediately next to the railroad line. He hoped in this way to lesson the city's impact on the national park.[9]

The building of the Hetch Hetchy system was a combination of public and private efforts. The first major city effort was to build a dam at Lake Eleanor. Out of the Hetch Hetchy Valley to the northwest snaked a rough gravel-and-rock wagon road to Lake Eleanor. Following an old trail, in 1916 a city crew had used generous amounts of dynamite to blast a road up the steep canyon and then cut the road to the lake, some 12 miles away. The city always assumed that Lake Eleanor and Cherry Creek would be tapped as part of the water system, but there was a reason for immediately developing Lake Eleanor. Both at Hetch Hetchy and at the city headquarters in Groveland, contractors had to have electricity, and a dam at Eleanor Lake and a turbine placed downstream on the combined Eleanor and Cherry creeks would provide it. By August 1917 "trains" of trucks, six in a row, wound their way up the treacherous, switchback road. They made three trips every 24 hours, hauling the cement necessary to construct the attractive, multiple-arch 1,260-foot-long dam. By late spring of 1918 workers had finished the dam, catching the spring runoff for power production. Kerosene lamps could be cast aside as electricity flowed up and down the river, from Groveland to the Hetch Hetchy site.[10] Electric power, necessary to move the project forward, was now assured. Soon the city engineers would begin plans for the much larger powerhouse at Moccasin Creek.

While city workers constructed the Lake Eleanor Dam, sawyers working for a private contractor cleared the Hetch Hetchy Valley. O'Shaughnessy paid A. J. Reeder $30,050 to strip the valley pine and oak trees, as well as riparian vegetation, with the understanding that all the trees would become either sawlogs or cordwood.[11] Denuding the valley did not cause O'Shaughnessy even a second's reflection. It was all business. One wonders if Secretary of the Interior Franklin Lane, establishing in May 1918 that "national parks must be maintained in absolutely unimpaired form," ever reflected on the paradox of the ongoing destruction of the Hetch Hetchy Valley and his principle of park management.[12] Deep in the Sierra Nevada mountains, loggers and hundreds of laborers did exactly what Lane proclaimed against.

Completing the infrastructure was the 58-mile Hetch Hetchy Railroad. The city needed to find a way to transport men, huge amounts of equipment, and thousands of tons of cement to the dam site. The city engineer's office considered constructing a road capable of handling large trucks, but in the end they decided that a railroad offered the most economical solution. Because San Francisco had no desire to own or operate a railroad, O'Shaughnessy tried to interest the Southern Pacific, the Santa Fe, and even the small Sierra Railroad in building the line and providing the service. But the prospect of building a line that would have little use after completion of the dam did not appeal to the companies. In the end San Francisco reluctantly decided that there was no alternative. The city would have to build and operate the railroad. It would be, according to *Railway Age,* "the first steam railway of any considerable extent to be built and also operated by a municipality."[13] By the summer of 1914 city crews picked the general route and surveyed the roadbed. In October 1915 the construction job went to Frederick S. Rolandi of San Francisco, the low bidder at $1,543,080. The contract called for Rolandi, a former railroad engineer who had turned to construction, to grade and lay track on 58 miles of line and then lay rails over the nine miles of new road into the valley, a total of 67.6 miles of

FIGURE 15. The valley had a tragic, naked look after sawyers had stripped it of tress and much of its vegetation. To the city's credit San Francisco did clear the valley, something that the Bureau of Reclamation often neglected, leaving graveyards of trees along reservoir banks. *Courtesy of the SFPUC.*

track.[14] Although everything seemed in place, construction by Rolandi would be delayed, simply because the money was not available.

The railroad, when finally completed in 1918, became the artery that gave life to the project. Connected in the west to the Sierra Railroad, a local line, the Hetch Hetchy Railroad wended its way through 27 miles of Sierra foothills to project headquarters in Groveland. From Groveland the track stretched to Moccasin Creek and then up the long, twisting 17-mile Priest grade, then along a ridge above the Tuolumne River, before terminating at the Hetch Hetchy station. Although the line was not officially designated as a common carrier until July 1918, the tough Shay steam engines moved vast amounts of materials to Groveland and then on to the end of the track, the flat cars loaded with cement bags, earth-moving equipment, and all the supplies necessary for a large workforce. In time the railroad carried approximately 300,000 tons of cement for the dam, the aqueduct, and various construction projects. Trucks, gasoline and electric engines, turbines, lumber, and vast amounts of iron and steel construction materials, all arrived by rail. Often the returning trains hauled logs to sawmills for private lumber companies, providing a little income for the city.

Besides carrying vast amounts of materials, the Hetch Hetchy Railroad provided the most convenient transportation for workers, engineers, city supervisors, and various dignitaries. Even prior to the powerful locomotives, O'Shaughnessy converted gasoline cars, buses, and trucks for use on the rails. Tiny "Sheffield" cars, fitted with railroad tires stretched over the wheel rims, were the railroad's official vehicles. They were fast, zipping along over the tracks at speeds as high as 50 miles per hour. They transported city officials and visiting dignitaries, but more often provided quick trips for construction supervisors. In a medical emergency, the Sheffield cars could transport an injured worker to the Groveland hospital. Most workers rode the rails in a streetcarlike bus that held up to 20 persons. But also one might see a converted Cadillac, one or two Packard trucks, or a Pierce-Arrow car at Groveland or at various work camps along the route. During the winter months, when snow blocked all roads and when turning out a steam train for just a few people was too expensive, these railroad vehicles were the transportation of choice.[15]

By 1919 human agency had transformed the Hetch Hetchy Valley and the Tuolumne Canyon below. Engineers and laborers had completed the Lake Eleanor dam and the Hetch Hetchy Railroad. Loggers had cleared the valley of its rich natural covering, electricians had built power units and strung wires, and carpenters had erected the many necessary buildings. Everything seemed in place for the major project: a dam that would submerge the valley under 200 to 300 feet of water.

Although O'Shaughnessy and his city engineers designed the Hetch Hetchy dam, it was the Utah Construction Company that won the bid and built it. The company consisted of William H. Wattis and his older brother, Edmund. The two cigar-smoking Mormons left their ranch and hometown of Uinta at a young age to work as teamsters for the Great Northern and Canadian Pacific railways. When they learned what they needed to know, they moved on. Soon the boys contracted to do their own railroad tunneling and grading. The Panic of 1893 destroyed their business and sent them back to the family ranch. But in 1900 they ventured out again. This time they were successful, perhaps keeping in mind the verse in the Book of Mormon "Inasmuch as ye keep the commandments of God ye shall prosper in the land."[16]

Now in their early fifties, the two hardworking, risk-taking brothers were interested in more than grading and paving the nine-mile road they had built into the Hetch Hetchy Valley. They hoped this small job would lead to a big one. The Wattises wanted to get into the dam-building business, and the proposed Hetch Hetchy Dam would offer them that chance. They knew that they could work with Michael O'Shaughnessy. The brothers and O'Shaughnessy, although not of the same religious persuasion, shared an ambitious, entrepreneurial, hard-driving attitude that spelled success. They believed that calluses on the hands were greater marks of accomplishment than three-piece suits and oak-paneled corporate boardrooms. Presumably, O'Shaughnessy encouraged William and Edmund to take the plunge after successful construction of the road. The brothers managed to bid under $7 million for the great gravity arch dam on the Tuolumne River. They won. It was their first attempt at dam building, the beginning of an adventure that would culminate in the building of Hoover Dam.[17]

FIGURE 16. City Engineer Michael O'Shaughnessy encouraged visitation to Hetch Hetchy Valley, especially after the city completed the railroad in 1918. Here a guide gives a group of San Franciscans an idea of what will happen in the next year or two. *Courtesy of the Yosemite National Park Archives and Library.*

When the city engineering office prepared the specifications for the dam, utility trumped Freeman's illustrations of ornate Doric gateways with ivy covering the upper dam and a lovely esplanade with motorists and tourists enjoying themselves. These architectural frills, as O'Shaughnessy labeled them, were too extravagant.[18] Beauty, for O'Shaughnessy, would be measured by utility and structural integrity. The 92-page bid document provided extensive details about the dam's specifications and measurements. Primarily designed by City Engineer R. P. McIntosh, it would be 298 feet thick at its base (almost 100 yards of concrete), tapering to a height of 341 feet, 227 feet above the valley floor. McIntosh designed the dam so that it could be raised and the water capacity increased as demand required. When full, the reservoir would back up the length of the valley, about 7.5 miles. The massive concrete dam would be the largest civil engineering project on the West Coast and purportedly the largest dam in the world.[19]

The bidding document spelled out the responsibilities of the city and of the successful bidder. For instance, the city would provide electricity, but the contractor would purchase that power. The city would provide a hospital at the site, but each month the contractor would pay $1 per employee for the upkeep of the facility. The city would receive $3 per cord for the firewood it stacked in the valley, but the contractor's employees would load and transport the wood. Bond forms were included, but no specific amounts were entered.[20]

Since the city had constructed the diversion tunnel and the cofferdam, the company could get right to work. Once again, financial problems arose. The city issued bonds but at such a low interest rate that bankers and investors ignored them. To get the project under way, the Wattis brothers had to purchase a couple million dollars' worth of bonds, an arrangement between contractor and contractee that was quite common in the world of large-scale bidding. Essentially the Utah Construction Company took out a loan to buy San Francisco bonds at face value, although the securities might be discounted as much as 10 percent. Once the city had the money, the project could begin and Utah Construction could draw on the funds it had provided. All of this worked to the city's advantage and seemed to be the price a contractor had to pay to win a bid. And, of course, a smart contractor understood the system and built the bond expense into his bid.

Utah Construction first excavated the dam site to granite bedrock. This required steam shovel operators to remove hundreds of tons of sand, gravel, and boulders from the riverbed. Down and down went the shovels, searching for bedrock. Below 65 feet the granite walls became too narrow and the steam shovels had to be retired. The company brought in derricks, which could

FIGURE 17. Digging equipment had to excavate 113 feet below the river surface to find bedrock. All this debris was hauled into the valley by rail and was spread by steam shovel. *Courtesy of the SFPUC.*

reach further down and then deliver "skips" of debris to waiting railroad cars. When loaded, the narrow-gauge cars were pulled by locomotive into the valley, where they were dumped and steam shovels spread the contents across the floor. After digging for months, the Utah Construction foreman "would periodically say, 'Well, it's deep enough.' Chief O'Shaughnessy would hurry up from San Francisco, take a look, then say, 'No. Go deeper.'"[21]

Finally, at 118 feet, there was no more debris. The workers could now sandblast and roughen smooth bedrock surfaces to provide adhesion, and think about pouring concrete. In preparation, Utah Construction built a double-track, narrow-gauge railroad 1.5 miles up the valley to what would become the gravel quarry. "Dinkies," small four-wheel locomotives, pulled a string of filled four-yard-long dump cars to the dam site, where Ransome mixers combined gravel, aggregate, and cement. Cement-weighing machines assured that the engineers had the proper formula. Describing the work to the Commonwealth Club of San Francisco, O'Shaughnessy called it "cyclopean masonry," consisting of poured concrete with boulders weighing more than one ton deposited throughout the mass.[22]

Cranes plopped these huge rocks into the forms, but four hoists lifted the concrete mixture to the top of a huge holding structure, 340 feet high. From that height an operator delivered the concrete by gravity pipe to whichever form workers were filling. This structure was impressive for visitors, but the cable span across the canyon drew more comments. Stretched across 900 feet of space at a height of nearly 400 feet, the two-and-one-fourth-inch cable carried 15 tons and was used to lower machinery as large and heavy as a narrow-gauge locomotive to the valley floor. Employees and often guests rode the cable in a basket. It was an adrenaline-producing ride that one reporter described as "the nearest approach to airplaning [*sic*]." There were harrowing experiences, but the most frightening was that of George Warren. One evening, timekeeper Warren lost his balance and fell from the basket, plunging 125 feet to the earth. When horrified workers hurried down to recover his body, they were amazed to find him alive. He had the spectacular good fortune to fall into a four-foot-deep pool of water. Not all of the men were so fortunate. During the more than three years it took to complete the dam, 17 Utah Construction workers died in accidents.[23]

Such a death toll might seem high, but during the pouring of the cement, a period of about two years, approximately 500 workers were on the job. They worked in three shifts, around the clock. The dam site was a constant swirl of activity, much of it seemingly chaotic, with trucks, conveyor belts, trains, hoists, and whistles all adding to the tumultuous din of construction noise. What a change for a valley once praised for its solitude! Hundreds of lights and spotlights transformed the night. A reporter returning on the steep Lake Eleanor road wrote that from his vantage point the dam site appeared like a fairyland, with hundreds of lights chasing away the darkness. "The crusher plant, the hoists and conveyors, the dinky trains—everything twinkles with lights. And above the dam, the great flood lights [shone] . . . as though it were broad daylight."[24]

It is difficult to judge the ability of O'Shaughnessy and his corps of engineers to supervise the hundreds of workers, particularly since we have only the management viewpoint. No doubt there were disgruntled laborers, but when San Franciscan Elford Eddy, a journalist for the *San Francisco Call,* visited the site, he found committed men. His stories emphasized the positive, such as the one at the Hetch Hetchy mess hall: After a fine meal, one of O'Shaughnessy's guests asked if the workers received the same food. Just then a group of Utah Construction Company workers arrived, and Eddy observed that they were, indeed, enjoying the same fare, as well as "the good cigars the 'Chief' had passed around."[25] Another journalist wrote of the co-operative spirit that seemed to prevail among the workers: "When a new man comes on the job, they show him every trick they know, help him do his share of the work with both speed and safety." The mentoring system extended beyond work, so that the new worker would be shown the "best method of washing his clothes," and the holes "where the best and most trout are found in the mountain streams, and how to catch them." And if he wished to augment his wages, "they show him the riffles, where, if he likes, he can go down and pan for gold." Another visitor praised the morale and commitment of the men, who apparently believed that the Hetch Hetchy project was more than a job. They had pride in their work. Waxing eloquent, he announced that no worker "will have a finer monument to his memory, a more valuable testimonial of his honesty, his ability and his public zeal."[26]

O'Shaughnessy often visited the construction sites to check progress, and though a tough boss, he exhibited an egalitarian streak that endeared him to many laborers. One journalist watched a worker approach O'Shaughnessy, throw his arm around his shoulder, and say "Now, Chief, I tell ye man to man."[27] If the man was working hard, O'Shaughnessy would listen hard. L. B. Cheminant, one of the Chief's valued engineers, told of a tunnel cave-in that trapped Tim Regan. His buddies forced through a pipe to provide him air. O'Shaughnessy spoke through the pipe. Regan asked, "How long?" "Oh ain['t but] two or three days," O'Shaughnessy replied. "Am I still drawing my pay?" Tim came back. "No," said O'Shaughnessy with a grin, "you're fired."[28] One can be sure that Regan got his pay and more.

Considering the difficulty of providing food and shelter in an isolated mountain environment, San Francisco provided well for its employees and those of the Utah Construction Company. Certainly the men knew they had not joined a country club, but the kitchen staffs at the various camps took pride in preparing wholesome and sufficient meals. Electricity, good water, and plenty of wood in the winter characterized the various construction camps—amenities often lacking on other work projects. Workers were satisfied with conditions. Hetch Hetchy, with a construction period of nearly 20 years, faced no strikes, although there were a few labor organizers in the camps. A good number of the workers, particularly those who toiled in tunneling, worked for O'Shaughnessy from the beginning in 1915 through to the completion in 1934.

Wages varied somewhat according to a worker's skill and the danger involved. An average might be $4.50 a day, with $1.25 subtracted for room and board. However, a laborer had every chance of earning a bonus of 12 to 30

FIGURE 18. Various vehicles were used for transportation, some fitted to run on the railroad or, like this Packard truck, to operate on rough, muddy roads. "Chief" O'Shaughnessy is in the passenger seat. *Courtesy of the SFPUC.*

percent of wages as determined by the progress and effort of his work gang. For instance, the city engineers set the base for tunneling at 300 feet per day, and if the three crews working 24-hour days accomplished more, then a bonus could be expected. Working at least six days a week, a laborer could accumulate a stake if he so wished.

Although the city tried to avoid labor strife, O'Shaughnessy was not altogether successful. In the summer of 1922 unidentified men began to show up in the camps. Strike notices appeared, tacked on trees and surreptitiously distributed, asking workers to stay away from all camps until management met their demands. The strike committee called for better wages, food, lodging, and for various other concessions.[29] The city did not budge, and the strike never got off the ground. City engineers and foremen fired some workers, claiming that those released were lazy rather than militant. But the reasons for work slowdowns were not always easy to determine, especially

FIGURE 19. During the height of dam construction in 1921 and 1922, the workers labored around the clock, creating dramatic night views of the construction site. *Courtesy of the SFPUC.*

for city supervisors and engineers who had little desire to talk with workers or entertain their complaints. Engineer N. A. Eckart reported in 1922 that "the effect of the I.W.W. strike and the tendency to slow up some part of the work were still apparent and it will be necessary to fire a number of men who had come on the job apparently with no intention of doing any hard work."[30] How many International Workers of the World organizers lived in the camps was unknown, but there were enough that City Engineer C. R. Rankin wrote the superintendent of Yosemite that he wished ranger assistance to prevent labor agitators from stirring up trouble.[31] Also, O'Shaughnessy noted two "Carpenters' Union" delegates who demanded that the city raise the scale from $6 to $9. O'Shaughnessy brushed aside their demand, and the strike effort soon evaporated. Throughout construction the Chief had no sympathy for labor agitators, especially those of IWW persuasion. He was not alone in the early 1920s, for as one historian has noted, "no state pursued its [labor] radicals more remorselessly that did California during this period."[32]

Many laborers saved a portion of their paychecks, but others spent whatever they made. In the work camps and on the job, San Francisco enforced strict rules. Workers caught drinking could expect to be fired with no fuss or appeal. Often they were "escorted" to one of the San Joaquin Valley cities to look for a new job. At one camp, named the Second Garrotte, an off-duty miner stumbled onto a bootlegger's hideout, took all the jugs he could carry back to camp, and offered the hooch to his buddies. Soon most of the men of the morning shift and the swing shift were drunk. In Groveland Supervisor Charlie Baird received a distress phone call. In short order he arrived unannounced at the Second Garrotte. He piled the besotted miners into two trucks, drove them to Stockton, and paid them off in cash.[33]

Yet in Groveland, project headquarters, no such rigid policies existed. The town loved to entertain the construction workers, especially when they arrived with paychecks in hand. It was wide open, and even though the American nation had begun its experiment with prohibition, there was never a shortage of liquor, either the real stuff or locally distilled "jackass whiskey." The day after payday, according to one witness, "drunks would be lying all over the streets of Groveland." The men did not limit their entertainment to drinking. Such establishments as the Groveland Hotel featured gambling in the lobby and rooms in the back known as the "bull pen." Prostitution was widespread, to the point that much later a Groveland resident believed that there "were only about two places that weren't at one time or another whorehouses—the post office and the church."[34]

San Francisco authorities turned a blind eye and a deaf ear to Groveland's rather unsavory reputation. Although the town served as the headquarters for the whole Hetch Hetchy project, the bars, hotels, bawdy houses, and the sheriff's office had no official connection with San Francisco. A visit to Groveland could blow off steam, and later the broke worker was happy to retreat back to the mountains. If a drinking worker stayed in the area, no harm was done, and he would be back on the job on Monday and not off in Sonora, Stockton, or San Francisco.[35] The Chief appeared to worry about Groveland, particularly when seasoned workers were involved. In 1922 John Burkland ended up in the hospital as a result of "bootleg stimulant celebrating his birthday." O'Shaughnessy was also concerned that "there were 15 or 16 men intoxicated riding up the line from Groveland in the bus." He advised that "steps should be taken" to check the use of bootleg in Groveland.[36] But the Chief was a decisive man, and if he had meant to reform the town, he would have done so.

Later, as workers laid the aqueduct pipe across the more populated San Joaquin Valley, O'Shaughnessy was less sympathetic to carousing. The Chief received an anonymous letter informing him that at certain construction camps fully half of his crews were drunk all the time. The letter singled out the Indian Creek camp as one of prostitution and endless supplies of bootleg whiskey. "Every pay day prostitutes are summoned from San Francisco and Livermore and Pleasanton, they pitch their tents outside of camp and pass as tourists." The informant warned, "We will watch and see, the press will certainly get this information if nothing is done."[37] The tough boss quickly investigated and corrected the situation. O'Shaughnessy never allowed the Hetch Hetchy project to be sullied by improper conduct by any of his men, especially if the San Francisco press, always looking for a story, might be lurking. Aside from the press, O'Shaughnessy understood the danger of liquor on the job. Speaking to the San Francisco Down Town Association in 1925, he said that from the beginning of construction, "there has been no booze on the job, and I am not a Volstead [prohibition] man, but booze and derricks do not work together. You have to have one or the other, and we have the other."[38]

By the spring of 1923 the Utah Construction workers were nearing completion of the massive concrete dam. It was time for the dedication. In a tribute to the talent and hard work of the Chief, in April the San Francisco supervisors voted to name the structure O'Shaughnessy Dam. On July 7, 1923, a large group of visitors from the towns of Priest, Sonora, Groveland, and of course, San Francisco assembled on the lawn in front of the bungalow where the chief engineer often stayed. It was a warm afternoon, although a breeze swept up the canyon, whirling clouds of spray from the spillway water. City employees clustered chairs under trees before the cabin for the occasion. Shade would be needed, for it was hot and the number of politicians in attendance guaranteed a long afternoon of speeches.[39]

First, William Wattis of the Utah Construction Company formally turned over the dam to Timothy Reardon of the San Francisco Board of Public Works. Then Mayor Rolph addressed the crowd, praising the Chief and patting himself on the back for having the wisdom to hire him. Then, turning toward O'Shaughnessy, he mentioned that an admirer had suggested a fund of $100,000 be raised "as a testimonial of appreciation from a grateful people." There is no record that the generous gesture ever came to fruition. James Phelan spoke next. Although the voters of California had recently turned him out of the Senate along with most Democrats, he was in good spirits. He reviewed the long struggle, expressing an undisguised scorn for the politicians, secretaries of interior, and environmentalists who fought against the city.

O'Shaughnessy came last and spoke longest. He introduced his engineers and reviewed the history of the project and the dimensions of the great dam. But in his concluding remark he returned to the old issue, seemingly tucked away in the recesses of his mind:

> The O'Shaughnessy Dam and Hetch Hetchy reservoir stand as a refutation to so-called "Nature lovers," who opposed its construction. Here, spreading for seven miles up the valley, lies a placid lake which is destined to become a magnet to all real nature lovers. Caught between two mountains and trapped by the barrier of stone and steel, it is a sight that would warm the hearts of those who delight in natural beauty. In the distance two waterfalls, like two immaculate ribbons, streak down the side of the canyon and are reflected in the huge man-made mirror. None have more accurately expressed the scene than the caption writer of a San Francisco newspaper, who placed the words over a photograph of the dam and reservoir: "Where Beauty and Utility Wedded."[40]

FIGURE 20. The 1923 dedication of the O'Shaughnessy Dam featured cowboys, park officials, bathing beauties, Indian maidens, and a child. *Courtesy of the SFPUC.*

FIGURE 21. Workers completed the dam in 1923, and here it is in operation. In 1938 San Francisco raised the dam and doubled the holding capacity of the reservoir, so that the current structure does not resemble the one in this photo. *Courtesy of the SFPUC.*

With this peroration O'Shaughnessy, Phelan, Rolph, and the city politicians buried any lingering doubts regarding the transformation of the valley. On that sunny day the participants would have appreciated a journalist who wrote of Huntington Lake, to the south of Hetch Hetchy, that "nature must have designed this spot to cradle a lake. As she made the sheltered valleys and wide alluvial plains, and invited man to make them fruitful by his labors, so she hollowed out these great spaces in the granite hills and waited patiently for the men of wide vision and vast resources to wall up the narrow outlets with their titanic buttresses of steel and concrete."[41] So it seemed to those in attendance at Hetch Hetchy: A city of imperial vision had, through wealth, human toil, and technology, achieved a perfect example of the technological sublime—the marriage of creative engineering and nature. The reservoir, following the contours of the valley, had waited patiently for creation, and now the engineers of San Francisco had fathered a "second nature" through transformation of a remarkable valley.[42]

It was a great day for Michael O'Shaughnessy. He must have been amused at Berkeley mayor Beverly Hodghead's comment that the Chief would probably not need a tombstone for a great many years, "but you may some time, and I think this is a pretty substantial one."[43] His engineering friend Clifford Holland wrote a congratulatory note from New York. Milt Young, a colleague from Ohio, summed up how O'Shaughnessy was feeling when he wrote that "your satisfaction in accomplishing this great work in the face of legal obstacles, governmental red tape, petty politicians and the Lord knows what, must be greater by far than any pecuniary reward or words of praise."[44]

As O'Shaughnessy accepted such praise, he was well aware that 23 miles downstream an equally important project was nearing completion: the Moccasin Creek Power House. By early 1922 city carpenters had erected a tent city adjacent to the site to provide rude housing for 300 men. Soon workers poured a massive foundation and a large powerhouse took shape. As with the dam site, the men worked round the clock, seven days a week, through the hot summer of 1922 and on into the winter. Although architectural perfection is not usually associated with powerhouses, this 285-foot-long building designed by architect H. A. Minton came close. It featured fifteen Spanish mission–style arches on each side and a tile roof, giving it a California flare. From up the hillside four penstocks dropped 1,300 feet, assuring a powerful head of water that in turn rotated and whirled eight Pelton waterwheels and four 20,000-kilowatt generators.

By June of 1923 workers finished the construction and the installation of the generating machinery. Also, city engineers tested the penstocks and the turbines. Everything seemed ready for full voltage. Only an accident can make one realize the fearful power of water if released where it should not be. On the morning of June 30, an employee testing the system forgot to open the bypass valve, creating a surge of tremendous pressure. Both the number 1 and the number 4 penstocks burst, sending a wall of water down the mountain, washing away soil, gravel, and rock as if it were confectionary sugar. The water and debris slammed into the tent city, and a number of wood frame houses, destroying everything in their path. The powerhouse sustained little permanent damage, but the basement filled with 15 feet of mud and debris. The cleanup took six weeks. Finally, in August, production commenced, activating the 98-mile electrical transmission line to San Francisco Bay.[45]

For Bay Area residents observing the progress of the Hetch Hetchy project, there were questions. Why had the city completed the long transmission line to the Bay Area, whereas it had not even begun the aqueduct and pipe? Had not the voters in 1909 authorized $45 million to bring Sierra Nevada water to the city? Why was O'Shaughnessy totally ignoring water? The public, believing that the purpose of the Hetch Hetchy project was to bring pure mountain water to the city, was confused. There was an explanation. The Raker Act required the city to use only water provided by the Spring Valley Water Company until such time as the city purchased the system or more water was needed. In 1924 the city had not yet succeeded in purchasing Spring Valley, and the company continued to provide for the city's water needs. In short, there was no reason to transport Hetch Hetchy water to the city, and to defer aqueduct construction would minimize the city's financial burden.[46]

Spring Valley had become respectable. Under William Bourn's leadership the company showed no signs of capitulating to a municipal system. In 1910 Bourn contracted with architect Willis Polk to build a water temple and place it at the point where the waters of the Alameda watershed gathered. Patterned after the water temple in Tivoli, Italy, the classic structure served no real function, but it replaced an old rough wooden shed with a classic temple that bespoke a glorious past and a hopeful future. One could imagine Isadora Duncan dancing at its base in a gauzy, flowing Roman toga. It was also a reminder of the hegemony of San Francisco, spreading its water tentacles into the hinterlands. Above the pillars and around the tower, stone masons etched an inscription: I WILL MAKE THE WILDERNESS A POOL OF WATER. THE STREAMS WHEREOF SHALL MAKE GLAD THE CITY. Surely the great temple celebrated the value of water, and Bourn felt that his company could, indeed, "make glad the city."[47] To accomplish that task, Spring Valley continued to expand its system to meet demand. In 1913 the company built the Calaveras Dam. It suffered a near catastrophic failure in 1918, but still Spring Valley adequately met San Francisco's needs. City residents did not seem inclined to pass the bonds necessary for purchase. On three occasions bond issues failed. There was much speculation about the failure, but Bourn could assume that his company would continue to provide the city's water for some years to come. Unless the city came to its senses and purchased Spring Valley, it could be decades before Hetch Hetchy water could flow through San Francisco faucets.

O'Shaughnessy and his Irish and Swedish laborers were, of course, committed to completion of the project. But what would the city do with a flow of 400 million gallons per day (GPD)? This was the figure Freeman considered sufficient for the year 2000! In 1915 the city and its participating suburbs consumed about 133 million GPD. The only logical strategy was to slow down construction of the aqueduct.

The hope of O'Shaughnessy, Mayor Rolph, and the city supervisors was to create a regional water system. At the time of the Raker Act debates, there was every indication that Hetch Hetchy water would flow through household pipes in Oakland, Berkeley, and Richmond. These three cities participated fully in the Washington hearings and debates. They shared information and put together studies demonstrating their joint needs. Oakland representative Joseph R. Knowland made a long, impassioned speech in 1913 promising cooperation. "What I desire to impress upon this House," preached Knowland, ". . . is the fact that this grant is not for a single city, but for the entire bay region." He cited a population statistic of 829,955 for the six counties adjacent to San Francisco Bay—35 percent of the entire state and rapidly increasing. To those who feared that San Francisco might not consent to participation by the East Bay cities, he maintained that such speculation was groundless.[48] Knowland's regional approach surely helped in the passage of the Raker Act, and when the Senate tallied the votes, at least two attorneys representing the East Bay were in the gallery seated alongside the San Francisco delegation.

FIGURE 22. Schoolgirls visit the Sunol Water Temple circa 1912. The temple became a favorite place for picnics and hikes. *Courtesy of Rebecca Douglas.*

The communities had every reason to cooperate, fulfilling the promises of Representative Knowland. However, there was no urgency for a decision. East Bay people knew that not until the 1920s or later would the aqueduct be completed. Periodically San Francisco representatives made overtures, but essentially Oakland took the issue "under advisement." In retrospect city attorneys should have come to an agreement by 1915. As in San Francisco, a private water company served Oakland and Berkeley, and residents were every bit as discontent as their neighbors across the bay. In 1920 O'Shaughnessy wrote an article on Hetch Hetchy for a Bay Area magazine, concluding that it was logical that the East Bay cities join with San Francisco, sharing the system and profiting from San Francisco's experience with "clever manipulators" who might lead the cities down a crooked path. When that time came, San Francisco would be generous, for "the prosperity of neighboring communities is inevitably linked with the advancement of our own city."[49]

Time was on the side of the East Bay cities. Although their water was expensive, they also understood that the longer they waited, the more desperate San Francisco would become. The East Bay cities realized they could drive a hard bargain. In early 1923 Oakland and Berkeley made the first step toward independence by voting to organize their own municipal utility district. The first question that the East Bay Municipal Utility District addressed was whether to build their own system or join with San Francisco. EBMUD secretary J. H. Kimball began a correspondence with O'Shaughnessy, first querying the city engineer whether it would be 15 years before Hetch Hetchy water was available. O'Shaughnessy assured him that it could be available within four years if $30 million or $32 million became available. The message was clear: A paying partner could accelerate the project and get the water flowing.

However, Berkeley and Oakland leaders were not prepared to commit. They preferred to gather information and debate, rather than sign agreements. In February 1924 Marston Campbell, president of the EBMUD, wrote Mayor Rolph, asking that the city staff respond to 24 questions regarding the Hetch Hetchy system and the East Bay cities' possible participation.[50]

Of course the San Francisco city engineer's office provided detailed answers, although O'Shaughnessy was not pleased. To John Freeman the Chief wrote that Campbell used to be "a hardware clerk in Honolulu" and "is seeking to flash himself into our Hetch Hetchy problem at this time." O'Shaughnessy confided that it was not surprising Campbell was creating problems for "he is an old student of Dr. Manson" and was raising construction issues "which I thought were permanently scrapped 12 years ago." Freeman answered sympathetically, reiterating his belief that "the whole district and separate communities should be supplied at wholesale by a metropolitan water system, almost precisely like that in Boston."[51]

However, San Francisco was not Boston. Whatever O'Shaughnessy may have thought of the EBMUD president, he was moving the East Bay cities inexorably away from San Francisco. In March 1924 O'Shaughnessy took the ferryboat across the bay to make his pitch to the Soroptimist Club of Alameda County. He believed that since Oakland and San Francisco were the two largest cities in the United States suffering the high rates of private water companies, they must work together to create a regional system. In the near term, he explained, the East Bay cities must face the problem of selecting a future water source. Although he refrained from saying so, he believed that anyone with his wits about him would choose Hetch Hetchy. He had worked out a formula for cost sharing. The East Bay, with a population of 350,000, would contribute $27 million, while San Francisco, with a population of 600,000, would contribute $54 million. The East Bay cities would receive 133 million gallons per day, enough to take care of a population of 1,650,000 people. He mentioned that there could be prejudices and local difficulties to overcome, "but by approaching the subject in a good spirit all those objections could be overruled."[52]

"A good spirit" did not prevail. At the same time O'Shaughnessy was attempting to entice the EBMUD, the district hired consulting engineer Stephen Kieffer to investigate an independent system. Kieffer surveyed the possibilities of the Mokelumne River and took options on the necessary land and water rights. He also brought in the consulting firm of Alvord, Burdick & Howson, Chicago-based engineers. Their report was damaging to San Francisco's hopes. The engineers noted that an independent system would be free of actions by San Francisco and that the proposed water source would be 40 miles nearer than the Hetch Hetchy system. Also, the Mokelumne project would not be saddled with electric power controversies. The consultants pointed out other features, most significant that the Mokelumne River project could be progressively built, expenditures made as the water was needed. In that way, Oakland, Berkeley, and Richmond would not suffer extensive loss on unused construction.[53] The consultants might have been compiling a list of Hetch Hetchy project weaknesses. As a member of the San Francisco negotiating committee, O'Shaughnessy continued to work, but the opportunity seemed to be slipping away, and there was nothing the city could do. Friends of San Francisco formed the East Bay Hetch Hetchy League with the intent of electing a new EBMUD board, but it was all to no avail, especially when an engineering committee made up of

Arthur Davis, George W. Goethals, and William Mulholland recommended the Mokelumne River project.[54]

On November 4, 1924, East Bay residents went to the polls and ratified a $39 million bond issue to build an independent water system. Key to the system would be the Mokelumne River and a high dam at Lancha Plana, storing sufficient water to supply an 81-mile aqueduct to the East Bay.[55] By early 1929 the district completed the Pardee Dam, one of the highest in the world at that time, and finished the Mokelumne aqueduct. The first water deliveries of Sierra Nevada water arrived in late June 1929.[56] Ironically, the Mokelumne River source had been suggested by the Army Corps of Engineers in 1911 for San Francisco but had been rejected by a city obsessed with the Hetch Hetchy Valley and the Tuolumne River. Freeman and O'Shaughnessy's idea of a regional water system would go by the wayside, as the East Bay cities threw off the dominance of San Francisco to develop their independent water system.

More than water influenced the East Bay cities' decision to go alone. The leaders of Oakland saw no reason to perpetuate a subservient position. Oakland had the better location as a port city to serve the inland cities by road or rail. It was pushing toward annexation of neighboring communities, creating a "Greater Oakland." San Francisco had very little to offer, and its dream of creating regional hegemony through a water system did not win friends. It would take more than San Francisco could offer to bring the recalcitrant child into a regional family in which Oakland's independence might be tested. Why not declare independence, twisting the tail of the San Francisco lion? In a sense, the Hetch Hetchy water system reaped a basketful of ill feelings. The *Sacramento Union* put it well: "San Francisco has pursued for many years a policy of belittling Oakland, yet wonders now why the big city across the bay should object to the submergence of its identity by consolidation."[57]

Oakland's refusal to be dragged into the quagmire of San Francisco politics proved sensible. San Francisco leaders were constantly squabbling, and the high cost of Hetch Hetchy was often the subject. By 1924 the supervisors, Mayor Rolph, the city attorney, O'Shaughnessy, and many others were bitterly fighting over the legality of selling electric power from the Moccasin Creek plant to the Pacific Gas and Electric Company. Equally vexing was the refusal of city voters to allow purchase of the Spring Valley water system. It seemed that the city just hoped to wear out problems, rather than solve them. There was plenty of bad blood at City Hall.

In retrospect, criticism could be expected. The Hetch Hetchy project had been under construction for 10 years, and still there was no water. If the city sold electric power to the Pacific Gas and Electric Company, it would be in defiance of federal law. By 1924 the project seemed like a perfect mess. Major Kendrick, a member of the San Francisco Advisory Water Committee, spelled it out: "We have spent $45,000,000 and we have got to spend $33,000,000 more to get water. After we have done that . . . we can't use a gallon of that water until we have used up the entire resources of the Spring Valley [Peninsula watershed, Alameda watershed, and Calaveras Dam]." Kendrick estimated that unless something was done, it would be 20 years before Hetch Hetchy water could be used, 10 at the minimum. Another $38 million would be required to purchase Spring Valley. To conform to the Raker Act, the city would have to purchase the electrical infrastructure from Pacific Gas and Electric for $35 million.[58] Kendrick failed to mention that the National Park Service was adamant that the city must live up to its Raker Act agreement to fund roads and trails in and around the dam. The *San Francisco Bulletin* noted that if all the public projects were undertaken, the city would incur a bonded indebtedness of $296 million, approximately 46 percent of the value of all taxable property.[59] The city's financial problems were daunting. More and more, Hetch Hetchy seemed less like a reservoir and more like a bottomless pit, consuming vast amounts of the taxpayers' money. In many ways, the Freeman plan, magnificent as it was, placed a significant burden on a city whose ambition outstripped its capacity. San Francisco had placed itself in a perilous economic situation, hanging out on a half-cut limb with no safety net from the East Bay cities or anyone else.

Mucking is a word used in tunnel construction to mean the removal of debris as the tunnel goes further into the earth. Muckers have dirty jobs, often working with wet, oozing soils that tax a worker's fortitude. More common in our lexicon is *muckraker,* referring to journalists who expose scandals and generally rake through news dung heaps for rumor, either true or false. Both muckers and muckrakers were much in evidence in the final 10 years of the Hetch Hetchy project.

As noted, O'Shaughnessy's critics found ample material for second-guessing, particularly with regard to his decision to blast a 29-mile tunnel from the Central Valley to near the San Francisco Bay. Civil engineer John Freeman had designed the tunnel and the gravity system, but there were plenty of naysayers who believed pumping the water over the mountains was the more sensible route. It would be less expensive, and by 1929 the East Bay Municipal Utility District had successfully pumped water into the Oakland region. Yet O'Shaughnessy stayed with the Coast Range Tunnel, a project that would employ almost 2,000 men at the height of construction. Such a long tunnel had never before been attempted, and the difficulties of working through rock and sinking fresh-air shafts were daunting indeed. Injuries and deaths were not uncommon. One July night in 1930 the project supervisor, Buddy Ryan, got a phone call. There had been a bad accident. Rushing to the site, Ryan found that 12 men had died instantly from a methane gas explosion—probably caused by one of the men illicitly lighting a cigarette. Work stopped. An investigation ensued, the state of California enforced new regulations, and the work continued.

The muckrakers had plenty of work to do, as well. Ironically, in 1930 the city of San Francisco became dependent on the newly constructed EBMUD system. Four dry winters had depleted the reservoirs of the Spring Valley Water Company. No Hetch Hetchy water was yet available. In desperation San Francisco contracted with the EBMUD for water to be delivered through a 12.5-mile emergency pipe. The city contracted for the water and the pipeline, paying $1 million from the Hetch Hetchy account to pay for it.[60] Journalists subjected O'Shaughnessy and his staff to a good deal of abuse, as well as humor. Thousands of San Franciscans must have shaken their heads in wonder at being dependent on the East Bay cities for water.

Perhaps such ironies finally loosened the pocketbooks of San Francisco voters. In a remarkable flurry of economic activity, voters approved a $24 million bond issue in 1928 and another for $6.5 million in 1932 to finish the Hetch Hetchy project. When the city issued the bonds, however, the interest rate was again too low to draw many takers. In a remarkable instance of civic duty, A. P. Giannini, founder and president of the Bank of Italy (later the Bank of America), picked up millions of dollars of the bonds, thus allowing the water-power project to move forward. City officials, including O'Shaughnessy, also prevailed on tunnel and ditch workers to take city bonds in lieu of cash. Most did, although participation likely stemmed from the desperate need to keep a job during the deepening Depression, rather than loyalty to the project.[61]

Probably the most remarkable reversal came in 1928, when voters approved a bond issue to buy out the Spring Valley Water Company for $41 million. Primary-stock owner William Bourn, so long at odds with the city of San Francisco, was in ill health and confined to a wheelchair on his marvelous FiLoLi estate. He wanted to clear up financial matters, and thus he offered Spring Valley at what he considered a bargain price. On February 1, 1929, voters of the city and county of San Francisco agreed. Given the hard times of the 1930s, William Bourn and the Spring Valley stockholders did very well.[62] By March 1930 the San Francisco staff moved into the Spring Valley building on Mason Street. The struggle to acquire a municipally owned water system, begun in 1877, was finally over.

By 1932 the Hetch Hetchy project moved toward completion. Out in the

San Joaquin Valley, Youdall Construction Company workers trenched and then laid more than 47 miles of pipe. When the workers finished the 8-foot-deep trench and laid and covered the pipe, they held an intriguing ceremony. With the Oakdale town citizens in attendance, 33 workers celebrated by placing their fedora hats in a big pile and burning them.[63] Presumably the hats had done their duty by shading faces from the hot valley sun. Meanwhile Charlie Shea and his Pacific Bridge Company successfully laid the pipe under the San Joaquin River. Over at the Tesla Portal, muckers and drillers continued their work, and by 1934 they completed the coastal tunnel and connected the Hetch Hetchy pipeline to that of Spring Valley.

In the meantime, a French stone mason was working on a Roman renaissance temple just south of Crystal Springs Lake. As a child Albert Bernasconi worked in his father's stonework business in Annecy, France. As a youth he studied architectural ornamentation in Milan and architecture at l'Ecole des Beaux Arts in Paris. He moved to San Francisco in 1911 and soon found employment with John Galen Howard, chief architect for the University of California in Berkeley. He worked on the San Francisco City Hall, the War Memorial Opera House, the Pacific Gas and Electric Building, and Grace Cathedral. Now he was building the Pulgas Water Temple, commemorating the arrival of Sierra Nevada water from 167 miles away. He worked on the ornamentation of other buildings, but the temple was special. Later in his rather broken English, he would say, "This is one I don't forget. I built the temple, myself." In 1934 it was time for a dedication.[64]

The October 28, 1934, dedication date of the Hetch Hetchy system had a melancholy aspect, however. Just three weeks before the release of the waters, the man who built the system died. Michael O'Shaughnessy had been in good health and had written to a friend that his large dog forced him to walk a mile each day. But a heart attack suddenly ended his life. As the *San Francisco Call* put it, his death, just days before the final completion of a system "to which he devoted the dreams and efforts of 20 years," was a tragedy. The editorial noted that the newspaper had had its differences, but O'Shaughnessy's "mistakes were matched by achievements. . . . On everything he did was the mark of the sound and thorough craftsman in the engineering art. As a personality, he dominated all about him and his passing leaves San Francisco poorer and less interesting by one salient and colorful individual."[65]

As at the dedication of the O'Shaughnessy Dam, the Chief's name was on everyone's lips, but this time solemnly and with bowed heads. Nevertheless, October 28, 1934, was a day for celebration as thousands of San Franciscans gathered in a pastoral meadow behind Woodside to watch the water arrive—

FIGURE 23. The Pulgas Water Temple and Reflecting Pool as it looked in 1938. *Courtesy of the SFPUC.*

first a trickle, then a roar—at the temple and then depart down a concrete chute to the waiting Crystal Springs Lake. Mayor Angelo Rossi and Lewis F. Byington, president of the Public Utilities Commission, made speeches. Supervisor Jesse Colman gave a tribute to O'Shaughnessy. Senators Daniel Murphy and Hiram Johnson attended, as did the widows of two men so instrumental in the project, Mrs. John E. Raker and Mrs. William Kent. The main address, carried over a national radio hookup, came from Harold Ickes, Roosevelt's secretary of the interior. He noted in his diary that he gave a talk on conservation, "with, of course, appropriate references to the occasion that we were celebrating."[66]

It was a perfect fall day—sunny, yet cool. The crowd milled about, fingering their programs, which featured a nude Venus veiled in mountains and clouds pouring water from a large vessel on the city of San Francisco. They listened occasionally to speeches that added the builders of Hetch Hetchy to the mission padres, explorers, pioneers, and other dreamers and doers who had transformed California.[67] The center of the celebration was Bernasconi's 60-foot-high temple. Some listened to the roar of the arrival of 34 million gallons of water per day, while others looked upward to the biblical verse etched on the cornice. I GIVE WATERS IN THE WILDERNESS AND RIVERS IN THE DESERT TO GIVE DRINK TO MY PEOPLE. It seemed a twentieth-century version of manifest destiny—the uses of technology to further the hegemony of a great and dominant civilization. Those conversant with the Bible understood that for such a gift God expected loyalty and sacrifice. So also did San Francisco exact an environmental price and a human price for the arrival of the pure Sierra snowmelt water. A valley was gone and 89 lives had been lost in construction. But those who were present on that bracing Sunday afternoon were of one mind that the assurance of a bountiful water supply for the next 75 or 100 years was well worth the expense, the struggle, the invasion of a national park, the undoing of natural beauty, and the loss of human life. It had even been worth lawsuits, failure of bond issues, political discord, difficult engineering problems, the shortage of men and materials during the war, and the antipathy of many farmers, politicians, and nature lovers. There had been many moments when the people of the Bay Area wondered if the Hetch Hetchy project would ever be finished. Did it represent progress? Would the bad be quickly forgotten and only the good remembered? On that Sunday afternoon they celebrated the good.

THERE HAVE BEEN many changes and additions to the Hetch Hetchy system since its dedication in 1934. Workers added aqueduct lines, engineers designed and built new dams, and electrical engineers reconstructed old hydropower units and put together new ones. Such changes were to be expected, for the demands on the Hetch Hetchy system far exceed those of 70 years ago. After all, Hetch Hetchy would be part of the continual challenge in California of redistributing water "from areas of surplus to areas of deficiency."[68]

O'Shaughnessy and chief designer R. P. McIntosh had planned the expansion of their dam from the beginning. It would have been done in 1923, but money was unavailable and the need was not evident. However, in 1935 city officials were determined to move ahead, particularly because the storage advantages were so impressive. Leslie Stocker, who replaced O'Shaughnessy as chief civil engineer, made compilations showing that once constructed, the new dam would increase the depth of the reservoir from 220 feet to 306. However, the reservoir capacity would jump from 67 billion gallons of water to 117 billion gallons. But the expansion was tied to increased hydropower production, not water. It was for this reason, reiterated Stocker, that in 1935 the city decided to build the final section of the O'Shaughnessy Dam.[69]

On April 8, 1935, San Francisco awarded the Transbay Construction Company a contract for $3,219,265. The company represented a combination of five contracting concerns, of which three had been members of Six Companies, Inc., the firm formed to build the monumental Hoover Dam. In many ways San Francisco's timing was perfect. The Six Companies had just completed construction of Hoover Dam and were looking for work. So were millions of unemployed Americans. Taking advantage of the depths of the Depression, San Francisco could drive a hard bargain. The city could even get help from the state of California and Roosevelt's New Deal government programs.[70]

In the intervening years the Hetch Hetchy Railroad right-of-way had deteriorated. When the State of California asked the city if it could use 500 or 600 men for a public works project, Mayor Rossi knew just what to do with them. By the summer of 1934, the men had replaced 33,000 ties and repaired much equipment to make the railroad serviceable. After completion of the upgrade, many of the men moved onto related work with the Works Projects Administration. Later the Public Works Administration would assist the city with construction of new power lines. The participation by the WPA and the PWA was the first time that San Francisco had received direct aid from the federal government.

With experienced contractors just off the Hoover Dam project, everything went smoothly. The engineers did have to ensure that the new concrete would adhere to the old, which had been cooling, shrinking, and aging for 12 to 15 years. The construction supervisors resolved the problem by mixing more cement into the concrete mix, creating a richer formula. They then artificially cooled the new concrete as it cured. By 1938 the contractors had finished enlarging the dam, and once again it was time for a dedication.

City officials had hoped that President Roosevelt might dedicate the enlarged dam. He had done as much for Hoover Dam. But Roosevelt's secretary of the interior had become incensed over San Francisco's neglect of the power provisions of the Raker Act. When he signed off on the dam enlargement, it was with the stipulation that his approval should not be interpreted as his capitulation to the city. When the dam dedication came, Ickes was locked in a legal case with San Francisco that was headed for the Supreme Court. He let President Roosevelt know that his appearance might signal approval of the city's actions. With a low-key ceremony, the new dam received the blessings of the city in late 1938.[71]

Although not so visible as the Golden Gate Bridge or the San Francisco Bay Bridge—the two other great public works projects completed in the 1930s—the Hetch Hetchy system was far more expensive to build and the politics exceedingly more distasteful. Yet in the end all three stand today as testaments to the San Francisco Bay Area's coming of age.

However, the story of Hetch Hetchy did not end with the 1934 dedication. San Francisco found that although pure water proved beneficial, significant electricity coming from the Moccasin Creek Power Plant was more advantageous to the struggling city. Yet the way in which San Francisco chose to market this valuable product ran counter to the conditions of the Raker Act and also to the sensibilities of a determined secretary of the interior.

CHAPTER 8

The Power Controversy

[Congress would not have passed the Raker Act] "if we did not believe the 200,000 horsepower of electricity would forever emancipate San Francisco from the collar of the hydroelectric trust."

CONGRESSMAN WILLIAM KENT

THE HETCH HETCHY controversy should be seen nationally not only as a fight about water but also as a battle between the proponents of public power and those who favor private companies. Public-versus-private-power conflicts dominated the world of electricity for half a century. Some of the main advocates of San Francisco's position focused on the hydroelectric power issue, rather than water, national parks, or anything else. Senator George Norris, Congressman William Kent, and the influential Gifford Pinchot recognized that a peaceful social revolution was at hand. Electricity was coming of age, spreading throughout the country, changing the everyday lives of urban Americans. They believed that the benefits from Hetch Hetchy flowed less from water than from electricity. Who controlled this electrical revolution, and who profited from it? Norris, Kent, and Pinchot fought for public ownership of utilities throughout their political careers.[1] They were in the vanguard of a determined group of Progressives intent on keeping control of utilities in the hands of the people, or more specifically, municipal governments. They had already seen evidence of monopoly exacting unwarranted profits from customers. By 1907 Samuel Insull captured 20 power companies in the Chicago region, effectively combining them into Consolidated Edison. He maintained his monopoly not only through sharp business practices but also by bribery and favors from local politicians.[2] The Pacific Gas and Electric Company was moving in the same direction. Norris wished to stem the tide. Already the California private utilities were gaining a stranglehold on the urban areas. Southern California Edison expanded its capacity, and in the north, Pacific Gas and Electric dominated, spreading its network of wires to serve the needs of both small and large towns. For many in Northern California it seemed that the electrical revolution had spawned a hydra-headed monster determined to tyrannize the people with overpriced electricity, and armed with the political power to resist any sort of meaningful regulation.

PG&E and other companies centered their generating activity in the Sierra Nevada mountains. With no coal deposits available and the vast California oil fields as yet undiscovered, power companies looked to the numerous rivers rushing from the mountains; an ideal, renewable kinetic energy source. Combining its technical expertise and capital, PG&E built hydroelectric power plants and the lines to transmit the power to coastal communities. The company led the nation in developing "white coal" and also in the technology of long-distance transmission of power.[3]

By the time the Hetch Hetchy controversy reached the halls of Congress, legislators recognized the importance of electricity, although its benefits had not yet reached rural America. Most of the senators understood junior senator George Norris's plea to pass the Raker bill, since "its ultimate effect is going to reach away beyond the lives of any men who lives[*sic*]. . . . Pass this bill, sir, and millions of children yet unborn will live to raise their tiny hands and bless your memory." Employing an analogy his fellow senators grasped, the Nebraskan reported that 100,000 horses were now idle. "Harness them and put them to work—100,000 horses that do not have to be fed, that never will get tired or weary, that never die of old age, that will be working 100 years from now as they can work now, without tiring and without ceasing."[4]

Norris wanted passage of the act not only to encourage use of electricity but primarily to checkmate a monopoly and reverse the trend toward privately owned utility companies. If San Francisco gained the right to generate electricity, the city would be in a position to establish a municipal power system. Passage would not be easy, for Norris believed that private power companies, headed by the PG&E, bankrolled his congressional adversaries in a nationwide effort. The Sierra Club and the defenders of the valley had been duped: Naively doing the dirty work of the privately owned utility companies, they were sacrificial pawns while the king makers manipulated the game. Much later, Judson King, a confidant of both Norris and Gifford Pinchot and an administrator for the Tennessee Valley Authority, wrote that the power interests, hiding "behind well-meaning nature lovers, launched a nationwide campaign to defeat the Raker bill."[5] King offered no evidence for his charge.

When the Raker bill passed, San Francisco secured its electric power, but not in the way the Nebraska senator envisioned. In a sense, the private companies won out. Congress fully expected that the city of San Francisco would purchase and develop a municipally owned power network, ushering in a new era in which ownership of basic services, such as water and electricity, would reside with the people. Costs would be lowered and corruption stemmed. However, corruption or not, private entrepreneurs were well ahead of the city in developing a needed service that might also yield a profit. Both the Pacific Gas and Electric Company and the Great Western Power Company built extensive infrastructure systems in San Francisco by 1913, securing rights-of-way and capturing customers. Neither company expressed any interest in selling a profitable business to the city. Fulfilling the intent of the Raker Act would prove impossible. To this day, Pacific Gas and Electric owns the infrastructure in San Francisco and furnishes power to both residents and corporations. This story, often forgotten, is difficult to unravel, yet it is important, since the electricity the Hetch Hetchy system generates is as consequential as the waters it provides.

As with all Hetch Hetchy issues, the power story started with the Raker Act. William Kent was determined to make a strong statement to free the city from the for-profit water and power companies by adding section 6 to the bill. This section, true to Kent's Progressive principles, must be quoted verbatim, for it became the center of controversy for 70 years and is still at issue today:

> Sec. 6—That the grantee is prohibited from ever selling or letting to any corporation or individual, except a municipality or a municipal water district or irrigation district, the right to sell or sublet the water or the electric energy sold or given to it or him by the said grantee:
>
> *Provided*, That the rights hereby granted shall not be sold, assigned, or transferred to any private person, corporation, or association, and in case of any attempt to so sell, assign, transfer, or convey, this grant shall revert to the Government of the United States.[6]

There is a certain clarity to section 6. The language is lucid and the intent unmistakable. The city could not, under any circumstances, sell Hetch Hetchy power to a private utility, such as the Pacific Gas and Electric Company. Yet section 6 caused no end of legal briefs, manipulation, and bitter debate. The Raker Act's intent was that the city of San Francisco purchase the private companies and create municipally owned water and power systems. In 1928 city voters finally passed the bonds necessary to purchase the Spring Valley Water Company. However, the electorate consistently defeated bond elections to purchase the Pacific Gas and Electric Company. Writing his autobiography late in life, a frustrated George Norris devoted a whole chapter to the failure of his Hetch Hetchy concept. "Now, thirty years later—three full decades—the powerfully entrenched private interests which prevented San Franciscans from enjoying what belongs to them still thwart the express will of the American Congress, the clear-cut mandate of the federal courts, and the Department of the Interior, under both conservative and liberal administrations."[7] How could this have happened?

COMPLIANCE WITH SECTION 6 of the Raker Act became an issue in 1923 with the dedication of the O'Shaughnessy Dam and with substantial progress on the Moccasin Creek Power Plant. By early 1924 workers completed the city-owned transmission line to Newark, on the southeast end of San Francisco Bay. Soon an average annual output of 460 million kilowatt-hours of electric energy would arrive for distribution and consumption. Yet the city had not succeeded in purchasing the necessary infrastructure from Pacific Gas and Electric. Many said the leaders had not tried. The federal government, private power interests, and public power advocates all watched and wondered how the city would negotiate section 6.

City Engineer Michael O'Shaughnessy understood the situation better than any other San Franciscan. In 1916 he placed the highest priority on finishing the Moccasin Creek Power Plant so that he could "put the City's water to actual work," realizing that the power plant would generate significant revenue.[8] Yet how could the city profit from its power plant if section 6 prevented sale to PG&E, the owner of the San Francisco power infrastructure? In 1923 neither the city engineer nor any other city administrator had resolved the dilemma of peddling power without violating section 6. O'Shaughnessy finally concluded that the city must sell, at least temporarily, city-generated power to the Pacific Gas and Electric Company. He asked fellow engineer Herbert Hoover, then secretary of commerce, to do what he could with the Interior Department. "It may be rather a delicate matter for you to interfere in the operations of another department," but still O'Shaughnessy hoped that Hoover might encourage a "hands-off policy" until the city worked out its power problem.[9]

At home the problem escalated when Mayor James Rolph announced that he favored municipal ownership: "I do not believe the wholesaling of power to a corporation would be legal, as I read the Raker Act."[10] The city supervisors agreed. The *San Francisco Examiner,* controlled by the flamboyant William Randolph Hearst, backed up the mayor and the supervisors, chastising the city engineer. "We like your engineering, 'Chief' O'Shaughnessy, costly as it is. But when it comes to matters of public policy," crowed the paper, "we think that the people's elected representatives, and such able and civic-minded citizens as Senator [James] Phelan and his Advisory Committee, are the proper persons to judge."[11] At a meeting of the Commonwealth Club, O'Shaughnessy faced harsh criticism from Phelan, who after all, had spearheaded the city's drive for control of the Hetch Hetchy Valley. "The framers of the Raker Act had one thing in mind," emphasized the former senator, that "no invisible empire, no trust should ever seize from the people of San Francisco this great resource of Hetch Hetchy. You may have it, congress said, but you may not give it away." And William Kent argued that Congress would have never put up such a fight if it "did not believe that the 200,000 horsepower of electricity would forever emancipate San Francisco from the collar of the hydroelectric trust." Kent recalled that "Mr. O'Shaughnessy was with us in that fight and was ardent for straight municipal ownership and distribution. I don't know what's happened since."[12]

Mayor Rolph's Advisory Committee, headed by Phelan, faced four choices: (1) construction of a distribution system apart from PG&E, (2) purchase of the PG&E system, (3) sale of power at wholesale rates to PG&E, or (4) distribution of power through PG&E facilities and lines. Option 1 seemed both wasteful and expensive. Option 2 would take time, requiring an appraisal and the passage of a bond. Option 3 violated the Raker Act, section 6. Option 4 was the most sensible, and the committee recommended it. However, both PG&E, headed by Wigginton Creed, and Great Western, a smaller electric company under Mortimer Fleishhacker, were quite determined to prevent the city from getting into the power distribution business. They refused to negotiate.[13]

The private power loophole in the Raker Act became obvious. The act did not mandate the purchase of the Spring Valley Water Company and the Pacific Gas and Electric Company infrastructures. Since it was not imperative that the city purchase the infrastructure, both sides began a dance of acquisition. Rather than enter into serious negotiation, the newspapers and the people of San Francisco preferred to lambaste the private companies, contesting their value—refusing to go to arbitration—and in general making it difficult to pass the necessary bonds.

Given the situation, Mayor Rolph and the city supervisors edged toward defiance of the federal government. The company agreed to purchase all

FIGURE 24. The original Moccasin Creek Power House is on the left side of the photograph. The powerhouse generated a significant amount of electricity, but sale of the power to Pacific Gas and Electric was a controversial issue that ended up in the U.S. Supreme Court. *Courtesy of the SFPUC.*

Hetch Hetchy power, but it refused to serve as the city's distributor. PG&E executives knew they held the upper hand. They would not budge, and after a 14-hour supervisor's meeting in March of 1925, Rolph and 11 supervisors accepted the PG&E power purchase terms.[14] Few knowledgeable persons thought that such a sale was legal, but they all believed it was necessary. Simply put, the city had no way to distribute Hetch Hetchy power except to sell it, wheeling it into Pacific Gas and Electric Company lines. If nothing could be worked out, Hetch Hetchy power would have to be "dumped," unused and useless. After all, electric power, unlike natural gas, petroleum, or coal, cannot be stored. It is a situation of use it or lose it.

On July 1, 1925, San Francisco Board of Public Works member T. A. Reardon and PG&E's vice president, F.A Leach, signed the agreement whereby the company would purchase the entire Hetch Hetchy power output for approximately $2 million per year. The compact emphasized that this was a short-term partnership to precede San Francisco's arranging to purchase and operate its own distribution system. The contract could be terminated by either party with *one day's* written notice. The city negotiators surely anticipated an agreement of short duration, but they also hoped that the conditions would convince Secretary of the Interior Hubert Work that the city meant business and would soon purchase the PG&E infrastructure. In a sense, the agreement was tailored to impress the interior secretary, for he had the power to cancel the pact if "in his opinion the agreement violates any provisions of the laws of the United States in general, or the Raker Act in particular."[15]

In another step to impress the Interior Department, Mayor James Rolph headed a delegation that went to Washington and personally presented the contract to the secretary. Covering his bases, he also sent a long telegram to President Calvin Coolidge which emphasized that under the agreement $5,479 would come to the city each and every day, whereas at present the power was "Being Wasted And Waste Will Continue Indefinitely If Contract Not Be Executed." He urged the president to take up the matter with Interior Secretary Work. Also, both Rolph and O'Shaughnessy appealed to Commerce Secretary Herbert Hoover to intercede on behalf of the city. However, some leaders were disgusted by what they believed was a subterfuge. Gifford Pinchot and William Kent urged Secretary Work to schedule open hearings in San Francisco. They both devoutly believed that the will of Congress would be undermined if this deception went unchallenged. Kent was particularly annoyed, wiring the secretary that when the city accepted the terms of the Raker Act it "Entered Into Moral As Well As Legal Obligations Which It Should Be Forced To Carry Out In The Near Future."[16]

Arriving in Washington in the summer of 1925, Mayor Rolph and his San Francisco delegation hoped for a quick approval from Secretary Work. The Californians found the Department of the Interior in disarray, or at least unable to take a position. The *Washington Daily News* condensed the frustrating process and even added a dash of humor:

> When SF came to Washington with their plan to sell to PG&E, "in violation of the federal grant," the matter fell into the jurisdiction of Assistant Secretary of the Interior [John H.] Edwards.
>
> Edwards was besieged with arguments and protests on each side, and decided to do nothing until Secretary [Hubert] Work should return from a western trip.

> Work took a brief look at the controversy, decided his opinion should hinge on strictly legal aspects of the matter, and passed the whole thing over to Atty. Gen. Sargent.
>
> Sargent got his assistant going, and in the end stated to Work that "he did not care to give an opinion on the contract."
>
> So back came the dynamite, with the fuse still sizzling, to torment Work. And Work rounded the circle with nicety by announcing that nothing will be decided until Edwards returns from a western trip and advises him.
>
> All of which is a decided, if temporary, victory for the people of San Francisco.[17]

In effect, Secretary Work and his staff refused to make a decision on the legitimacy of the PG&E contract; but neither did he cancel it. Tired of such ambivalence, Mayor Rolph returned to San Francisco and settled the issue. He would not allow the indecision of the Interior Department to cost his city $5,479 a day. On August 14, 1925, the delivery of Moccasin Creek plant power commenced and electricity flowed through the PG&E lines into the city. It was a defiant act, one the *Los Angeles Record* labeled "contempt of Congress, the President's Cabinet and the Judiciary."[18] In retrospect the refusal of Secretary Work to uphold section 6 of the Raker Act was an error that may be charged to indecision, inertia, and bureaucratic bungling. September came and went with still no decision. Finally, in early October, Assistant Secretary Edwards announced that the city could not sell Hetch Hetchy power to others for resale. He was too late. Once the electric power flowed, it would be hard to stop.

Still, selling Hetch Hetchy power to Pacific Gas and Electric Company was not considered a final solution, but rather a stopgap measure, a temporary agreement until the California State Railroad Commission, the predecessor to the California Public Utilities Commission, determined a value for PG&E and the Great Western Power Company. The city could not ignore Edwards's decision. Neither could it ignore the verbal bashing by Congressman Lewis Cramton. The Michigan legislator aimed most of his barbs at the city's inaction in fulfilling its responsibilities in Yosemite National Park, but he also criticized the city for its violation of section 6. Cramton demanded action from Ray Lyman Wilbur, President Herbert Hoover's new secretary of the interior. Wilbur, past president of Stanford University and a person well versed on Hetch Hetchy, quickly addressed Cramton's criticism regarding Yosemite Park. About the 1925 power contract, however, he remained silent. The new secretary's sympathies lay with the private power companies. When it came to the Raker Act, he would prove to be a "loose constructionist."

In June of 1929 the Railway Commission rendered its valuation of the San Francisco properties and infrastructure held by PG&E and Great Western. Not surprisingly, both companies appealed, believing their properties were undervalued. The commission denied their appeals, and Mayor Rolph wrote Secretary Wilbur a long letter explaining the valuation and the importance of the upcoming bond vote, which he scheduled for August 26, 1930, the same day as the California state primary election. The total bond measure would be $68,115,000, but voters would cast separate votes for purchase of PG&E at $44.5 million, Great Western Power Company at $18.9 million, and $4.5 million for new transmission lines and power house construction. In spite of Mayor Rolph's hopes, the election proved a disaster for those who favored municipal ownership. On both the PG&E and the Great Western acquisitions the votes were approximately 25,000 for, but 63,000 against. The bond issue for transmission lines and power house construction also went down to defeat. It was a remarkable rejection of municipal power.[19] It seemed particularly significant since just two years earlier city voters had finally approved purchase of the Spring Valley Water Company by a four-to-one majority. Some blamed the two companies for influencing the election, while others noted the deepening economic depression. Citizens who followed the city's business felt that to approve the bonds would place a dangerous debt burden on the city and exhaust its bonding capacity. Whatever the reason for defeat, those who favored municipal power knew they had lost decisively, and to win would not be easy.

Not all city officials, to be sure, were committed to public power. Many, in fact, were quite satisfied with wholesaling Hetch Hetchy electricity. In a 1935 memorandum, Lewis F. Byington, president of the San Francisco Public Utilities Commission, noted that the partnership guaranteed $21 million in gross revenues to the city over the 10-year period, of which $16 million had been applied to reducing the heavy debt. Byington suggested that had these monies not reduced the debt, the voters "would have failed to vote the necessary appropriations for water called for in 1928, 1932, and 1933."[20] A voting pattern was emerging: San Franciscans would approve bonds to buy and to complete the Hetch Hetchy *water* system, but not to purchase the *power* distribution system.

In the meantime, a political change at the national level had motivated Lewis Byington's 1935 memorandum. Secretary Wilbur left office with the Hoover administration, and President Franklin Delano Roosevelt appointed Harold Ickes to head the Department of the Interior in 1933. Almost immediately the new secretary scrutinized the Raker Act and asked that the city explain why the 1925 power agreement was not in violation of the act. In his 13-page memo Byington set out the argument of the San Francisco Public Utilities Commission, claiming that the city had "reasonably complied" with section 6. The federal government, in Byington's view, could expect no more.[21] Ickes did not agree, and from 1934 until his resignation in 1945, the interior secretary proved to be a persistent, painful thorn in the city's side. In dealing with San Francisco he surely enhanced his nickname, "the old curmudgeon."

Nothing seemed to annoy Ickes more than shallow legal arguments aimed at obscuring the facts and avoiding the law. That is the way he viewed Byington's reasoning as well as the city's recalcitrance. It irked him that the city's refusal to comply with section 6 enhanced the profits of a private utility company. As a longtime advocate of public power, Ickes loved to twist the tail of any corporation that sought to increase profit at public expense. PG&E fell into that category. However, in his passion for public ownership, Ickes refused to acknowledge another power model gaining acceptance. This model was utility regulation through state commissions, a prerogative that appealed to many parties and stemmed the enthusiasm for municipal ownership.[22] In the 1920s and 1930s the California Railroad Commission regulated the rates that PG&E could charge its customers, and the commission kept profits in line. According to Byington's statistics, PG&E city rates were a little higher than those in the rural areas, but the return varied between 7 and 10 percent—not an outlandish profit.[23]

Besides being a stickler for the letter of the law—at least when it supported his purpose—Ickes was a public power ideologue, in many ways the antithesis of his predecessor. Loquacious and pugnacious, Ickes had a certain inelasticity that frustrated San Francisco officials and made cooperation difficult. In fact Ickes did not want cooperation so much as capitulation. By June 1933 he asked acting director of the National Park Service Arthur Demaray to review what had taken place between the city and the interior secretary's office. Demaray reported that the city had done nothing, even though the 1925 power agreement was temporary and was only in "reasonable compliance with its obligations for the time being."[24]

Irritated, Ickes continued his investigation into the 1925 contract, soliciting letters and briefs and even conducting an open hearing in which San Francisco and other interested parties presented their case. In August 1935 Ickes issued a 31-page brief. To no one's great surprise, the secretary concluded that if the 1925 agreement had ever been a temporary expedient, it was no longer. The city must carry out a clear and binding obligation under section 6. Never one to shy away from giving advice, Ickes suggested that the city amend its charter "to permit the issuance of revenue bonds which will require the approval of only a majority of those voting at the election."[25]

Actually, the supervisors agreed that it was time for a change. Until 1935 the city listed Hetch Hetchy power bonds as general obligation bonds, requiring a two-thirds vote for approval. Now the supervisors approved a charter amendment to list the bonds as revenue producing, requiring a simple majority. Passage of the amendment would open the way for the San Francisco Public Utilities Commission, which replaced the Board of Public Works in 1932, to move closer to purchase of the PG&E infrastructure. However, PG&E, the Chamber of Commerce, the real estate interests, and all the merchant associations vigorously opposed the amendment, stressing that if it passed, the public would lose "safeguards against extravagant spending by public officials."[26] Voters defeated the amendment, 72,780 to 50,804.

This defeat stunned Secretary Ickes. He could not understand San Francisco voters, especially since he had received numerous letters congratulating him for holding San Francisco's feet to the fire. These letters insinuated that corruption and greed had captured the city.[27] Whether PG&E controlled the city supervisors was problematic. The company advertised and lobbied, but there is no evidence of bribery or political chicanery. The city residents seemed equally divided between those who favored public power and those who opposed it. The majority may have supported public power, but according to one pundit, in the midst of a national depression and economic hard times, the people were reluctant to launch the city into the retail power business. Furthermore, purchase of the PG&E infrastructure would give employment to very few people. Solid reasons explained the voters' rejection of bond proposals.[28]

Angelo Rossi, who became mayor when voters elevated Rolph to the governorship of California, and the supervisors now faced the task of satisfying two formidable authorities. The voters of San Francisco had made it abundantly clear that they remained unconvinced by the public power advocates. On the other hand, Harold Ickes, who in a nationally broadcast speech heaped praise on the city at the 1934 Pulgas Water Temple ceremony, grew more determined that San Francisco would follow the letter of the Raker Act. In an attempt to mollify the secretary, San Francisco engineers, attorneys, and politicians created a number of "plans," one of which they hoped Ickes would find acceptable. Between 1936 and 1941 city officials submitted *nine* plans. In March 1936 Mayor Rossi and his staff journeyed to Washington to present Ickes four of them, each with a little different twist. Journalist Arthur Caylor noted that the upcoming gathering would be "a poker game rather than a power conference" with "Honest Harold." "The local boys hold four cards in the shape of four plans for municipal distribution of Hetch Hetchy electricity, and will back them with large numbers of white chips. But they'll hang onto their blue counters to use in connection with a wild ace which remains up the sleeve." Caylor thought Ickes both suspicious and informed.[29]

Caylor was quite right. In February Solicitor Nathan Margold had analyzed the four plans for his chief. None included outright purchase of the PG&E infrastructure; rather, all involved more complicated "wheeling" of power into PG&E transmission lines with proper payment. The "wild ace" plan was number 5, or what Margold called "4B." This one envisioned the delivery and sale of power to PG&E at its Newark line, then resale to the city at wholesale rates. However, Margold estimated that approximately 207,000 kilowatt-hours reverted to the company for sale at retail rates, a clear violation of the Raker Act.[30]

Ickes soon sent the mayor and his staff back to San Francisco to try again. Once more the supervisors backed a charter amendment that authorized $50 million for acquisition of the PG&E. This "Plan 7," according to Solicitor Margold, constituted compliance with the Raker Act. The election vote, on March 9, 1937, was closer: 77,514 no, 65,688 yes. A near miss did not impress Ickes. On March 15 Ickes issued an ultimatum that the Board of Supervisors must take action—what action he did not say—to comply with the Raker Act or he would ask the United States attorney general to bring court action in order to cancel the 1925 agreement. Rossi responded immediately, begging the secretary to delay any legal action and asking for another conference. Ickes wired back that his patience had worn thin and he saw no need of further discussion. In a memo to Ickes, Solicitor-General Frederic Kirgis mused that "apparently the Mayor was completely bewildered and disconcerted by the knowledge of the fact that conferences and delays would no longer be the regular order of things."[31]

On April 2, 1937, Ickes moved to the courts, requesting the Department of Justice to file suit to prevent further power sales to Pacific Gas and Electric Company. Events moved quickly. On April 11 Judge Michael J. Roche of the federal district court handed down a decision granting Ickes's wish for an injunction and declaring the 1925 agreement in violation of the Raker Act. However, Roche granted suspended enforcement for six months, until December 28, 1938. On October 17 San Francisco attorney John O'Toole filed an appeal to the Ninth Circuit Court of Appeals, tying up the whole matter in the courts. The appeal forced Judge Roche to grant a number of injunction extensions. Ickes felt bedeviled by the snail's pace of the proceedings, and in March he leaked to the press that the government might take over the Moccasin Creek Power Plant, citing the section 6 provision that if San Francisco violated the Raker Act, "this grant shall revert to the government of the United States."[32]

On September 13 the Ninth Circuit Court of Appeals overturned Judge Roche's decision, concluding that since the voters had refused to pass a bond act for purchase, San Francisco had no choice but to allow PG&E to act as "agent" for the city. Ickes immediately asked the Justice Department to appeal to the U.S. Supreme Court. On April 22, 1939, Supreme Court Justice Hugo Black delivered the opinion of the court, reversing the circuit court on the basis that "Congress clearly intended to require—as a condition of its grant—sale and distribution of Hetch Hetchy power exclusively by San Francisco and municipal agencies directly to consumers in the belief that consumers would thus be afforded power at cheap rates in competition with private power companies, particularly Pacific Gas & Electric Company." Therefore, Black wrote, the injunction was both appropriate and necessary. Representing an eight-to-one majority, Black's decision supported Ickes.[33]

While the case was working its way through the courts, the relationship between Secretary Ickes and Mayor Rossi and the supervisors deteriorated. Ickes, in charge of the massive federal Public Works Administration (PWA), was not above using the carrot of a federal grant to bring the city into compliance with the Raker Act. In early July 1938 a series of "monstrous telegrams"—as Ickes described them—between the two illustrated the growing frustration. Ickes asked Rossi if "The City Intends To Submit As Part Of Its Public Works Program The Proposal For Distribution Of Hetch Hetchy Power?" In response, the mayor argued that the city was doing everything in its authority. In May 1936 the supervisors submitted a proposal to Ickes for $53,127,000 for an electrical distribution center, of which $12,500,000 would be covered by a PWA grant. The city intended to put the $40,580,000 city contribution to a vote in the near future. It would be presented as a revenue bond, requiring only a majority vote. Ickes did not respond. Rossi accused the secretary of delaying tactics, claiming that evidently the application for $12.5 million reposed "In a Pigeonhole in your office." In high dudgeon, Rossi attacked what he considered Ickes's belligerent tone: "I cannot understand your stubborn and dictatorial attitude toward the good people of San Francisco, and particularly our unemployed." Although the mayor had been patient, Ickes had "not properly digested my telegrams," and the secretary was "sadly lacking in . . . knowledge of city and state legislature procedure in this which is one of the largest states in the union." Clearly the mayor could not be easily intimidated. Sensing, perhaps, that he had gone too far, on July 11, Ickes terminated the exchange:

> In View of What Appears to Be Your Continued Deliberate Intention to Misrepresent, Based on a Pretended Interpretation of My Language, Which Is Not Justified, I Can See No Point in Replying to Yours of July 9 Further Than This Acknowledgment.[34]

Ickes denied that he used any sort of intimidation with regard to Public Works Administration monies. He could point to his approval of PWA and WPA funds to assist the city in the project to raise the O'Shaughnessy Dam. Yet Mayor Rossi was not completely off the mark. In his July 16, 1938, diary entry Ickes reviewed the situation, noting that Rossi "pretended that I was threatening him with a refusal to approve any PWA project unless San Francisco decided to build the line. Frankly, this latter conclusion was not entirely an unjustified one. I suppose that there was an implied threat in my telegram, but as a matter of fact I had not decided that I would adopt such a course as that."[35]

We have no way of knowing whether Ickes did "pigeonhole" the San Francisco WPA application. However, if he did, he seriously undercut his own desire to have San Francisco comply with section 6 and bring the city the benefits of public power. On May 19, 1939, the supervisors once again submitted to the voters a charter amendment to issue $55 million in bonds for public power. The vote was decisive: 50,283 yes to 123,118 no.[36] If the bond amount requested had been for $40,580,000, with the understanding that the federal government would kick in another $12,500,000 in WPA funds, the vote count would have been different, although not enough to allow passage. On the other hand, Ickes accused Rossi and City Attorney O'Toole of failure to seek WPA funds. In 1941 the secretary suggested that had San Francisco sought to build its own electrical infrastructure (thus creating jobs), a WPA grant could have provided 45 percent of the cost. The city then could have funded the remaining 55 percent through a federal low-interest loan.[37]

The overwhelming vote against public power in 1939 indicated that virtually nothing Ickes, the mayor, or the supervisors could do would change the minds of San Francisco voters. Residents wanted a public water system, but they stood firm with the private power of PG&E. Why? Some believed that a publicly regulated private power company was more efficient than a bureaucratic municipally owned one. For many people the issue seemed to be one of political ideology. Numerous Republicans saw their vote as a protest against the New Deal. Elizabeth Cosby, writing to Ickes on behalf of the Women's City Club of San Francisco, complained that "the old, antiquated Horse and Buggy Raker Act is being used by autocratic, centralized government to circumvent the will of a free community." Cosby noted that "the vote has eight times defeated a political attempt to socialize the electric light business of our city."[38] PG&E advertising, aimed at undermining public power, influenced voters. Also, the company worked at establishing a broad ownership of the corporation's stock. Yet many San Franciscans preferred a regulated private monopoly over another government bureaucracy. They argued the efficiency of a profit-motivated company. Finally, a public bureaucracy appeared more economical only because, unlike a private company, it paid no taxes on its property and infrastructure. If one factored in the taxes PG&E paid to the city, the advantages of public power dwindled.[39]

Secretary Ickes must take some blame for San Francisco's refusal to endorse public power. It was a matter of personality. Many San Franciscans did not oppose public power as much as they did Ickes himself, whom they perceived as a bullying Washington bureaucrat sticking his nose in the city's business. In his determination to win, the secretary pulled no punches, using the courts, federal grants, and an uncompromising interpretation of section 6 to bring the city to its knees. At times Ickes seemed much less interested in civility and solutions than in confrontation, threats, revenge, and triumph. Although he denied it, the secretary seemed to have a personal vendetta against the leaders of San Francisco and even the voters—at least that was the opinion from Mayor Rossi's office and most of the local newspapers. The one political power that residents possessed was the vote, and they used it.

By 1940 the Hetch Hetchy power issue had become a confrontation between Ickes and the federal government, on one hand, and the people of San Francisco on the other. In mid-July a *San Francisco News* headline announced, "S. F. DELEGATION GETS READY TO FACE THAT TERRIBLE ICKES MAN ON SUBJECT OF HETCHY." The impression was that Ickes was immovable and impossible. After the delegation of Rossi, City Engineer Cahill, and City Attorney O'Toole returned from another Washington meeting empty-handed, the *San Francisco Examiner* chastised Ickes, a man who "stubbornly refused to permit SF to act economically and rationally." Journalist Arthur Caylor of the *News* suggested that since the delegation had returned with "a suitcase full of black eyes," stalling should be the new strategy. "The idea is simply that the election of [Wendell] Willkie would mean the end of Ickes."[40] However, most San Franciscans could not count on Franklin Roosevelt losing the election. Perhaps anticipating FDR's victory, City Attorney O'Toole penned a letter to Senator Hiram Johnson, inquiring if he might set up a meeting with President Roosevelt, thus bypassing the secretary of the interior.

The other solution, often bandied about, was to amend the Raker Act, clearing away the obstacle of section 6. When the November election returned Roosevelt, the California congressional delegation introduced House Resolution 5964, which simply said, "Be it enacted . . . that the words 'or the electric energy' be . . . stricken from Section 6." After 11 days of testimony the House Committee on Public Lands failed to move the bill forward. Issues that pitted public power against private utilities created controversy.[41] The bill died, and with it the hope of San Franciscans for a solution by amending section 6.

Meanwhile Ickes wrestled with what to do with his Supreme Court decision. There was no question that he would recommend that Justice Hugo Black's majority decision be enforced with an injunction. But when? In May 1941 Mayor Rossi, City Attorney John O'Toole, public utilities manager E. G. Cahill, supervisors, and citizens again journeyed to Washington. The party of 12 met with Ickes and four assistants in the interior conference room on the morning of May 21. From the onset the meeting went badly. Ickes expected that the delegation would have a plan, one that would finally meet the conditions of section 6. To his displeasure the delegation had none. He said, "You might just as well have saved your carfare." O'Toole pleaded that "all we are here for is to ask whether you will give us consideration and give us time to work out our plans." Cahill admitted he was working on a 20-year leasing arrangement with PG&E "which will completely satisfy every demand of the Raker Act." Ickes asked how long the lease agreement would take to complete, and Cahill opined that it could be signed in six months. Ickes doubted that estimate in view of San Francisco's track record "of delay, evasion and double-crossing." Ickes believed that "all you want is time enough until I am out of office."[42]

The secretary, rising to the occasion, lectured the delegation that his relationship with the city in 1933 had been friendly but then it had dawned on him "that cooperation with San Francisco was a one-handed implement, San Francisco holding the handle." Ickes pointedly asked O'Toole when the federal government had filed suit against the city. "April of 1936," came the answer.

> SECRETARY ICKES: That was 4 years ago and you started out with Judge Roach's [*sic*] opinion against you, so that any reasonable man might have concluded that he had only a 50-50 chance of winning the case in the Supreme Court, and during those 4 years what has the City of San Francisco done to get ready? All they did, apparently, was to come in to ask for more time, Mr. Cahill.

MR. O'TOOLE: I do not want to interrupt your trend of thought, but the record will show that during those 4 years upon two different occasions we presented to the people an amendment to the charter which, if passed, would have permitted the financing.

SECRETARY ICKES: Mr. O'Toole, you did not want those passed, and you did not raise a finger to cause them to be passed. The very profits that went to P.G.&E., some of it at any rate, went back into the election and went back to some of the organizations represented to help fight that amendment which was proposed.

The hearing continued in this vein, with Ickes commenting to O'Toole "that any man can be fooled once, but a man's a damn fool if he allows himself to be fooled the second time, and this isn't even the second time." After Ickes disposed of O'Toole, he quizzed Cahill regarding a lease of the San Francisco electrical infrastructure from PG&E. Cahill reluctantly agreed to try to negotiate a 20-year lease agreement for $50 million within 60 days, although he warned Ickes that "I do not think I can succeed."

SECRETARY ICKES: I wish you would go back with less of a defeated attitude.

MR. CAHILL: I do not have a defeated attitude. You are asking me to do something practically impossible.

SECRETARY ICKES: [Turning to Mayor Rossi] If I were you I would send somebody else over to Mr. Black's office [a PG&E negotiator].

The meeting, which focused on personalities more than issues, finally concluded with Representative Frank Havenner, who had worked for many years for some sort of resolution, asking that "we ought to abandon all hostility" and stop "throwing monkey wrenches in the whole machinery." The San Franciscans boarded the train with the understanding that Cahill would attempt to negotiate a lease agreement with PG&E within the next 60 days.[43]

In late August 1940 Mayor Rossi returned to Washington. Manager Cahill and City Attorney O'Toole stayed home, thus avoiding the secretary's wrath. However, at this particular meeting there were no tongue-lashings. Ickes, Rossi, and representatives Havenner and Richard Welch focused on support for a leasing understanding with PG&E. Ickes was intent on getting the San Francisco Chamber of Commerce, the newspapers, and various other business and civic groups behind another municipal bond election, or at least support of a leasing agreement. Everyone agreed with Mayor Rossi that "the Supreme Count has spoken and that presents a different case to the people."

With assurances that city officials and local business associations would help carry out the Supreme Court mandate, Ickes was willing to delay enforcement of the injunction until June 1, 1941.[44]

By late November 1940 the city had come to a tentative agreement with PG&E. The city would lease the company infrastructure for 20 years at a fixed charge of $4.9 million per annum. Amidst criticism from Ickes and others, Cahill claimed it was the best the city could do. After a number of discussions, Ickes finally declared that the proposed lease was not in the best interests of the city and not in compliance with the Raker Act.[45]

The secretary continued to wield the threat of an injunction, but he remained flexible. Rossi had indicated that in 1941 the city officials were in a better position to convince the people of the benefits of public ownership. Therefore, if the city scheduled another election, Ickes would recommend a stay of the injunction until June 30, 1942. He warned, however, "the ultimate limits of tolerance, beyond which a government official cannot go, have certainly been reached." Three questions must be answered: Did the city wish to own and operate its own electrical system? Did it want the benefits of public ownership, particularly lower rates? Did the citizens wish to comply with the Raker Act?

The secretary received his answers in November 1941, when again the citizens defeated the bond measure. To be sure, PG&E influenced the election by spending $62,000 on negative advertisements and strategically lowering rates; however, the voters were not going to support public power. Late in the month, the secretary informed Mayor Rossi that he had no choice but to enforce the injunction, with the federal government taking possession of O'Shaughnessy Dam and the Moccasin Creek Power Plant.[46]

As it happened, he did have a choice. The Japanese attack on Pearl Harbor, December 7, 1941, changed everything. For San Francisco it offered a reprieve from Ickes and a temporary solution to a very old problem. Suddenly the nation needed all available power, and to place an injunction on Hetch Hetchy hydropower would have been both unpatriotic and politically unwise. Ickes found a solution. He wedded wartime needs with Hetch Hetchy, terminating the 1925 power contract with PG&E and selling all available power to an aluminum smelting plant, built by the Defense Plant Corporation (part of the War Production Board) and located near Modesto, at Riverbank on the Stanislaus River. The city received $2,405,000 a year and was finally in compliance with the Raker Act. One San Francisco newspaper commented that "those whose favorite indoor pastime a few months ago was to think up new epithets to apply to Secretary Ickes would do well to recant now."[47]

Welcomed as it was, the aluminum plant proved to be a fleeting solution. Opened in early July 1943, the Defense Plant Corporation informed the Alcoa Company executives that the plant would close in August 1944. Aluminum was available elsewhere. Furthermore, a pollution damage suit for $580,000 (injury to crops, livestock, and people) gave notice to government officials that the plant had not been a good neighbor to the community.[48] In his annual message Mayor Roger D. Lapham addressed the difficulty, opening the old wound that some hoped had been closed:

> Last August without notice to the City, this aluminum plant was suddenly shut down and since then, the city has been delivering only a very nominal amount of power for upkeep of the plant. On June 28, 1944, Judge Roche of the U.S. Federal Court had given the city until August 28th to produce a plan . . . which would comply with the visions of the Raker Act.[49]

It was back to the drawing board for Ickes and Mayor Lapham. However, the drama included a new cast of characters. In early 1945 Ickes was still secretary, but Abe Fortas, his assistant, who later would be appointed to the U.S. Supreme Court, carried on the negotiations. Dion Holm, attorney for the San Francisco Public Utilities Commission, and James Turner, chief engineer, assisted Mayor Lapham. These players plus others could be found often at Washington's Carlton Hotel. They worked hard, and on May14, 1945, the city and PG&E entered a contract whereby the company wheeled Hetch Hetchy power from its Newark substation to various points in the city, to be used for the Market Street railroad, streetlights, and other city-owned facilities. The city paid PG&E in cash and on a kilowatt-hour basis and could not be paid in electric energy. Ickes and Fortas believed the contract was in reasonable compliance with the Raker Act.[50]

The May 1945 contract did not resolve the city's surplus energy problems. In Washington, representatives and attorneys for the Modesto Irrigation District (MID) and the Turlock Irrigation District (TID) were deeply involved in the discussions. The city and the two districts had compatible power needs. Essentially, the irrigation districts needed a power supplier, while San Francisco needed a market. On March 12, 1945, R.V. Meikle, TID's representative, and MID chief engineer Clifford Plummer signed a contract whereby the two districts would purchase *all* of the surplus Hetch Hetchy power. To ensure compliance with the Raker Act, the contract stipulated that neither district could resell the power to PG&E nor any other private utility.

The two irrigation districts had evolved into more than their names might imply. By 1940 they provided both water and power for their rural and urban customers. At the time that San Francisco built the O'Shaughnessy dam, the TID and MID combined their financial resources to construct the Don Pedro Dam on the lower Tuolumne River. The dam stored irrigation waters, but the powerhouse created valuable electricity. Initially, the Modesto Irrigation District Board of Directors intended to wholesale its 35 percent share of electricity to the Pacific Gas and Electric Company, the electricity provider for the town. However, the people of Modesto voted to have a publicly owned utility company. For some 20 years the Modesto Irrigation District and PG&E maintained competing electrical systems and fought for each customer. During that period the MID would have favored purchase of Hetch Hetchy power, but under the 1925 power agreement with PG&E, San Francisco had to deliver *all* of its power to the private company. The 1925 agreement not only worked well for PG&E in San Francisco but also gave the company considerable leverage against the MID.[51] Yet the majority of the people of Modesto kept fighting, determined to have public power. Finally on June 10, 1940, PG&E capitulated, selling its competing infrastructure for $325,000.[52]

The Turlock Irrigation District never suffered a fight for public power. The Don Pedro Dam contract called for two-thirds of the power to be delivered to TID, more than it would need for a number of years. While MID bought power at retail prices from PG&E, TID sold its surplus to the utility company at wholesale prices. Ironically, both irrigation districts accepted the doctrine of public ownership so central to the Raker Act, while San Francisco—where it was to apply—rejected the principle.

The March 1945 agreements between San Francisco and the irrigation districts worked well, and minor violations became less important in 1946, when Ickes bade farewell to the Interior Department. Oscar Chapman, the new secretary, had no ideological cross to bear; thus Ickes's "strict construction" of the Raker Act gave way to loose construction under Chapman and continued with the Republican ascendancy under President Eisenhower. The term *reasonable compliance* came back into vogue. In truth, however, with increased deliveries to the Modesto Irrigation District, the city was complying with the Raker Act. Quarterly reports from the San Francisco manager of public utilities, J. H. Turner, indicate that in 1951 and 1952 the city did not dump any power into PG&E lines.[53] Over the past 50 years, there have been investigations and efforts in San Francisco for municipal ownership. However, they have been brushed aside, and the dichotomy remains in San Francisco between a publicly owned water system and a private power company.

NEW ISSUES, HOWEVER, arose. As the need for electricity grew, San Francisco significantly increased the power output of the Hetch Hetchy system by developing the Tuolumne River and its tributaries below the Hetch Hetchy Reservoir. A dam on Lake Eleanor had been the first city project, but in 1949 the city, in conjunction with the Modesto and Turlock irrigation districts, signed a contract with the Army Corps of Engineers to develop Cherry Valley, to the west of Lake Eleanor. Workers constructed 26 miles of mountain road in 1950 and brought in heavy equipment. By 1956 city engineers completed the huge Cherry Dam, 330 feet high, 2,600 feet long, and 1,320 feet thick at its base. This facility sits outside Yosemite National Park. Six miles downstream on Cherry Creek, the city built the Dion R. Holm powerhouse. The massive dam and powerhouse had a dual purpose—to store water for power generation during peak periods, but also to provide the MID and TID with downstream water deliveries promised under the Raker Act.

The other major facility was the Kirkwood power plant. Engineers located the plant, finished in 1967, at the termination point of the Canyon Power Tunnel, which carried most of the Tuolumne River from the O'Shaughnessy Dam 11 miles down the canyon. With the Moccasin Creek, Dion, and Kirkwood powerhouses the city had the capacity to generate more than two billion kilowatt-hours of power per year. San Francisco consumed about 25 percent of that power, mostly through use in municipal buildings, street railways, and street lighting. The irrigation districts took the other 75 percent, although questions remain regarding the city's total compliance with section 6.[54]

One other cooperative project should be mentioned because it impacts the future use of the Hetch Hetchy system. In 1923 the two irrigation districts completed the Don Pedro Dam. However, by 1949 San Francisco, the Army Corps of Engineers, and the MID and TID combined their interests and monies to construct a massive new dam. The new Don Pedro Dam, dedicated by Mayor Joseph Alioto in 1971, dwarfed the old one, submerging it in 200 feet of water and creating a 165-mile shoreline. For its $45 million dollar investment, San Francisco received from the irrigation districts a Hetch Hetchy credit of 740,000 acre-feet of exchange-water storage space. Considering that the total capacity of both Lake Eleanor and Hetch Hetchy Reservoir approximated 366,000 acre-feet, this additional reserve gave the Hetch Hetchy system the kind of security against drought that it needed. The Tuolumne River, which averaged an annual runoff of approximately 1,800,000 acre-feet, always left the city unprotected against the inevitable dry years, since the irrigation districts were entitled to 1,090,000 acre-feet of that water. San Francisco needed about 453,000 acre-feet annually. However, as historian Alan Paterson has noted, such figures do not account for the wide variations in runoff from year to year. The huge reservoir behind the new Don Pedro Dam gave San Francisco abundant carryover storage for those lean years. The Don Pedro addition increased the city's high-mountain storage space to 1.4 million acre-feet, enough to provide a daily delivery capacity of 400 million gallons. Finally, San Francisco could realize the amount of water that John Freeman promised in 1913, although according to Patricia Martel, past general manager of the San Francisco Public Utilities Commission, the system only delivers a little over half that amount.[55]

Although the Raker Act forced cooperation in the development of the Tuolumne River, the 1945 power contract conditions were not satisfactory to the irrigation districts. Under this contract the city furnished all the electrical energy the two districts required. However, if the districts' needs exceeded the available supply, San Francisco purchased power from PG&E and then resold it to the districts, passing along the higher costs of supplemental purchases. As the demand continued to grow, the districts chafed under the arrangement, but there was little that San Francisco could do. In 1968 the two districts contracted with R. W. Beck and Associates to draw up their own power project for the lower Tuolumne River between Cherry Creek and the Don Pedro Reservoir. The consultants came up with a rather elaborate plan known as the Clavey–Wards Ferry project. They envisioned building two small reservoirs and 19 miles of 10-foot-in-diameter tunnel to transport much of the river to a powerhouse at Wards Ferry, just above the backwater of the Don Pedro Reservoir. The generating station would operate only at peak periods. It would be expensive to build, but the electricity would be valuable.

In April 1976 the Turlock Irrigation District and Modesto Irrigation District applied to the Federal Power Commission for a temporary permit. San Francisco joined the cause. The three partners did not envision serious challenges, but now they faced a new antagonist in the growing environmental movement. In the same year that R. W. Beck and Associates drew up the Clavey–Wards Ferry project, Congress passed the Wild and Scenic River Act, affording protection from such development for rivers that might qualify. Also in that year, 1968, kayakers Gerald Meral and Richard Sunderland, both members of the Sierra Club, ran the white water from Lumsden Bridge to Wards Ferry, the last remaining free-flowing section of the Tuolumne River. In the years to follow, more and more kayakers and rafters enjoyed the challenge of the river, and by 1972 commercial rafters began scheduling trips. These recreationists had created a viable commercial use of the lower river. Pressure to save the river emerged from environmental groups as well as commercial interests, resulting in a law in 1975 to provide for the studies needed to justify wild and scenic river designation.

Thus by the time the irrigation districts applied to the commission, the environmental interests were organized, led by the Tuolumne River Preservation Trust. By July 1977 the state, rafting organizations, and the Sierra Club prevailed upon the FPC (soon to be renamed the Federal Energy Regulatory Commission [FERC]) to withhold a permit until a detailed study could be made showing the impact on the Tuolumne Canyon of the Clavey–Ward Ferry project. Perhaps anticipating the passage of the National Environmental Protection Act of 1980, the environmental interveners simply wanted the two districts to produce an environmental impact statement. Beck did the best it could to minimize impacts, suggesting that the contractors use helicopters for much of the work, but the plan faced serious criticism. One fatal blow came from the United States Forest Service when it announced that no matter what FERC decided, the agency would not issue a special use permit.[56]

Despite the slim chances of gaining a permit, the districts fought on. Although San Francisco remained neutral, the districts found support from other water agencies, chambers of commerce, and labor unions. However, public hearings in Columbia, Modesto, Oakdale, and San Francisco revealed strong public sentiment against the project. The state of California, environmental organizations, and white-water rafting companies all testified to keep the river flowing. Friends of the River, a new environmental group that had cut its teeth in opposing the New Melones Dam on the Stanislaus River, presented a petition with 18,000 signatures. Nationally, *Readers' Digest* and the *New York Times* recounted the wonders of the Tuolumne River. The fight was essentially over when influential California senator Pete Wilson sided with the river interests. Finally on October 2, 1979, President Jimmy Carter asked Congress to make the Tuolumne River part of the wild and scenic river system.[57] The districts would have to look for other ways to generate electricity.

The fight to save the remaining Tuolumne River white water brings the story full circle, back to the original Hetch Hetchy Valley struggle. Both struggles were, in essence, to save natural places and remarkable scenery for recreational use. Both were about the river. Both involved respectable government agencies intent on doing the right thing, which they perceived as improving the lives of the people they represented. Both fights pitted the forces of engineering and technology against the natural river. Yet the outcomes were very different. The river below Hetch Hetchy Reservoir had been bulldozed, tamed, channeled to the point that virtually nothing natural remained. In 1913 the defenders of the valley had said "stop" and failed. Now the defenders of the river said "enough" and won. On the Tuolumne River the limits of human control had been reached. By 1980 it was evident that the Hetch Hetchy system and its civil engineers would no longer prevail and that further modifications would be fought by river protectors with intensity and tenacity—a legacy their forefathers had established some 70 years earlier.

CHAPTER 9

The Legacies of Hetch Hetchy

"If San Francisco cannot properly cooperate with its generous landlord, the relationship had better cease and the Federal Government resume exclusive use of the park area."

CONGRESSMAN LEWIS CRAMTON, 1927

"Hetch Hetchy—Once Is Too Often."

ROBERT CUTTER

DESPONDENCY DESCENDED ON John Muir and his fellow defenders of the Hetch Hetchy Valley when they lost their fight. But a glimmer of hope remained. The forces of technological progress and the power of San Francisco won control of the valley, but Americans had awakened to the vulnerability of scenery and the national parks. Writing to Henry Fairfield Osborn, Muir predicted that "wrong cannot last, soon or late it must fall back home to Hades, while some compensating good must surely follow."[1]

What might that "compensating good" be? First, the National Parks Act of 1916, often called the Organic Act, must be credited to those who fought for the valley. Also, the National Parks Association, created in 1917, drew its origins from Hetch Hetchy. An equally significant result of the fight came in the challenge to assumptions regarding the value of dams, particularly their erection in national parks, including two notable clashes in Yellowstone National Park. When new reclamation projects surfaced on the Colorado River in midcentury, an invigorated Sierra Club, harkening back to Hetch Hetchy, took on the dam builders with striking success. Perhaps most important was the way in which the controversy established boundaries between the claims of outside interests and the National Park Service. Succinctly, between 1915 and 1930 the city of San Francisco often acted as if one-third of Yosemite Park was its domain, based on the federal grant of the Hetch Hetchy Valley. San Francisco's aspirations drew heavy fire from Park Service administrators and congressional leaders, leading to a subservient position for the city and a statement of authority by Yosemite National Park. Finally, the legacy cannot always be measured, especially when it is an idea or commitment to the national parks. "Remember Hetch Hetchy" has never been a popular chant, yet the Yosemite case worked in more silent, deeper ways, influencing and inspiring dedication. It can be seen as the first environmental cause that attracted national support, one that should not be forgotten.

For San Francisco, the city gained a magnificent reservoir site and an unequaled water supply, of which it is justly proud. It also acquired the right to develop hydropower, a valuable and needed product. From the city's perspective the Hetch Hetchy agreement of 1930 demonstrated that a public agency and the National Park Service could work together, with each gaining what it needs. Essentially, a few areas of a park could be multifunctional, while most should be reserved for scenery. Hetch Hetchy sanctioned this dichotomy, and it provided the opening wedge in which other public or private organizations would work together with the Park Service toward a common end.

Finally, the Hetch Hetchy story has an intellectual legacy. Environmental activists and historians have perpetuated the myth that the idea of wilderness preservation rested at the heart of the defenders' case. There is another myth, one largely of my own making. When I began research, I expected to find sufficient evidence to state that the defenders of the valley used a "rights of nature" argument. However, my search of the voluminous documentation of Hetch Hetchy turned up no such evidence. Perhaps a few seeds of the wilderness preservation and "rights of nature" ideas can be discerned, but the full flower would not come until 30 years later.

THE DEFENDERS OF Hetch Hetchy discovered that guarding the national parks against a determined invader requires an organization to argue, lobby, and, if necessary, make sacrifices. San Francisco had hardly finished celebrating its victory when those who had opposed the project began working to make sure it would not happen again. How could the national parks and exceptional lands be protected? It was not enough to have a few environmental and mountaineering groups, almost always short of money, trying to defend against powerful forces. If the Hetch Hetchy fight accomplished anything, it magnified the need for a federal agency committed to the parks. The agency would administer parks, but more significant, define their meaning and mission and, when necessary, defend them. Such an agency would help to heal the weak and wounded system and give credence to Muir's hope that the fight for Hetch Hetchy had not been in vain. The 1916 National Parks Act partially fulfilled Muir's hope.[2]

However, passage of the National Parks Act was not a direct result of the Hetch Hetchy "steal." The effort was well under way before passage of the Raker Act. After 1909 the nature lovers fought on two fronts: one to save Hetch Hetchy, the other to protect the national parks. There was, of course, a decided overlap between the two objectives. When William Colby outlined three preservation objectives to save the valley, he included a fourth: to "lobby for a general bill to safeguard the National Parks."[3] Thus, the effort to retain the valley and to establish a National Park Service moved in concert. There was also an overlap in personnel committed to both ends. Perhaps most representative was J. Horace McFarland, sometimes called "the father of the National Park Service."[4] In 1910 McFarland, on one of his frequent trips to Washington, stopped by the Department of the Interior, seeking national park information, but he could find no desk, office, or even a responsible individual. He vented his frustration to his good friend Frederick Law Olmsted Jr. and did not hesitate to do the same to Secretary Richard Ballinger. Consequently, Ballinger invited McFarland to confer with his staff and to draw up a national parks bill.[5]

McFarland and Olmsted realized the need to empower a new agency based more on recreation and tourism and committed to preservation of historical objects and natural scenery. They wished to create a federal agency that would view land and water and mountains with a different mission than those of the Bureau of Reclamation and U.S. Forest Service, both of which emphasized the commodity value of natural resources. This proposed agency might provide a declaration of independence for land from capitalist enterprise and economic development. Under the National Parks Act of 1916, then, a small portion of the land would be managed "to conserve the scenery and the natural and historic objects and the wild life therein to provide for the enjoyment of the same in such manner and by such means as will leave them unimpaired for the enjoyment of future generations." The long and bitter debate over Hetch Hetchy influenced the shape of this mission statement, the touchstone for managing the parks and "the principal criterion against which preservation and use of national parks have ever since been judged."[6] McFarland would later comment that had a proper mission statement been on the books in 1910, the loss of Hetch Hetchy Valley could have been averted.[7]

Those who shepherded the national parks bill through Congress had been deeply involved with the Hetch Hetchy fight. Both sides seemed to mend their ideological disagreements. In December 1915 Representative John Raker introduced House Resolution 434, a bill to establish the National Park Service. William Kent became an official sponsor. Again, two determined supporters of San Francisco's claim to Hetch Hetchy sponsored a bill to help ensure that such an invasion would not happen again. Horace Albright recalled that the bill's advocates held continuous strategy sessions. Kent, Raker, Stephen Mather, Robert Yard, Robert B. Marshall, J. Horace McFarland, Richard Watrous, Gilbert Grosvenor, Enos Mills, William Colby, Fairfield Osborn, and others gathered for innumerable meetings in Kent's Georgetown home, Yard's apartment, the Cosmos Club, or the offices of the National Geographic Society.[8] With this dedicated group and the help of innumerable volunteers, the national parks bill moved quickly through Congress, and President Woodrow Wilson signed it into law, thus establishing the National Park Service, on August 25, 1916.[9]

With the exceptions of Grosvenor and Yard, every one of the men who worked on the national park bill had participated in the Hetch Hetchy struggle, and certainly not all had fought against San Francisco. Albright believed that Raker "always felt badly about his part in promoting Hetch Hetchy and hoped he could redeem himself by pushing through a national park service bill with his name attached." Raker's inconsistency bemused some of the legislators. "What was this?" they asked, according to Albright. "A fellow who helped destroy part of Yosemite is now mothering a national park bureau?"[10] The same might be said of Kent, although by the 1920s this strong advocate of public power questioned his own support of the Raker Act.

The fight over Hetch Hetchy not only illustrated the need for a park agency but also clarified that the U.S. Forest Service should not be that agency. Although Gifford Pinchot never mentioned Hetch Hetchy in his autobiography, *Breaking New Ground,* the controversy played an important role in limiting the Forest Service's jurisdiction. Had it not been for the controversy, Pinchot and the service might have succeeded in becoming the guardian of the national parks. Control of the parks had been Chief Forester Pinchot's intention from the time he succeeded in moving the Forest Reserves into the Department of Agriculture in 1905. In that year he wrote Robert Underwood Johnson that he would have nothing to do with the national parks "until they also are turned over by law to the Department of

Agriculture." In case Johnson should miss his message, he stated that "I am very strongly of opinion that the National Parks ought to be in the same hands as the Forest Reserves, and I hope something may be done in that direction at the next session of Congress."[11] Secretary of the Interior James Garfield endorsed the idea that the parks and forests were "practically the same."[12] The following year Pinchot, at the height of his administrative power, committed himself to the San Francisco cause, suggesting to Marsden Manson, shortly after the earthquake and fire, that San Francisco "make provision for a water supply from the Yosemite National Park." He would "stand ready to render any assistance which lies in my power."[13] Reading between the lines, Hetch Hetchy under Forest Service governance would privilege the city's interests over those less inclined to sacrifice the valley. Pinchot never wavered from his ambition, placing whatever barriers he could muster to defeat congressional bills creating a separate national park agency.

If Pinchot's primary purpose was to gain control of the national parks for the U.S. Forest Service, his Hetch Hetchy strategy proved flawed. His determination—some would call it stubbornness—often frustrated and, indeed, angered the defenders of the valley. In supporting San Francisco's claim, he alienated the most important conservationists responsible for the National Parks Act of 1916. Horace McFarland, William Colby, Muir, Allen Chamberlain, and Frederick Law Olmsted Jr. came to despair of changing Pinchot's dogmatic position regarding the valley. In eight years he had not budged, writing and testifying in 1913 that "the thing seems to me to be clear beyond the possibility of arguments. To put it baldly—the intermittent esthetic enjoyment of less than one percent is being balanced against the daily comfort and welfare of 99 percent." Such a strong Progressive social statement ultimately prevailed, and yet in stamping a utilitarian seal on Yosemite National Park, the idea of turning over *all* the parks to the Pinchot philosophy seemed absolutely repugnant. Pinchot and forester Henry Graves's argument that there was little difference in the management objectives of the national forests and of the national parks seemed incongruous to McFarland, who defined the national parks as "the nation's playgrounds" and the national forests as "the nation's woodlot." The chasm between the two definitions was so deep and wide that, in McFarland's opinion, it could not be closed. Pinchot's uncompromising stance on Hetch Hetchy was evidence enough that neither he nor his Forest Service could be entrusted with the national parks.[14] Gradually, particularly under the Wilson administration, Pinchot's and the Forest Service's unrelenting pressure to administer the parks lessened, and the idea of a new and separate agency took hold.

The Hetch Hetchy struggle also demonstrated that organizations and strong coalitions were essential for success. The long battle had been spearheaded by private environmental groups that consisted of dedicated volunteers willing to write letters, stuff envelops, and give of their organizing ability and occasionally their money. When William Colby determined that the Sierra Club was too local and splintered, he formed the Society for the Preservation of National Parks, which engaged a national constituency. After 1913 the society disappeared, but the idea did not. No sooner did Congress authorize the National Park Service than Mather and Albright recognized the need for a private support organization that would fund studies, define objectives, and take positions that a public agency would find difficult. The result was the National Parks Association (now the National Parks and Conservation Association), founded in 1917 and headed by Robert Sterling Yard, a New York journalist pressed into service by his friend Stephen Mather. Yard's personality was too acerbic and abrasive for government work; thus he joined with Charles D. Walcott and Henry B. F. Mcfarland to form this support organization.[15] The association fought to preserve parks as undisturbed natural refuges, defending them from the many interests that sought political gain or economic opportunity. This mission remains today.

In 1916 the new park agency had a difficult job. The Bureau of Reclamation had little respect for national park landscapes. In Glacier National Park, created in 1910, the enabling act allowed the Bureau of Reclamation to enter the park and develop and maintain reclamation projects. The bureau soon claimed its right, damming Shelbourne Lake within the park. Given free rein by Secretary Lane, the bureau set its sights on St. Mary's Lake, arguing that a dam would assist farmers in both Montana and Canada. Much later, the Army Corps of Engineers, not to be left out of the reclamation feast, proposed the Glacier View Dam, which would plug the north fork of the Flathead River and submerge some 20,000 acres of the most primitive lands within Glacier Park.[16] But by this time, the new National Park Service was active, and Park Director Newton Drury protested. It was just one of numerous attempts by irrigation and flood control interests to penetrate the parks. Enough was enough, for the loss of Shelbourne Lake had already offered an example of what should be avoided if American national parks were to remain showplaces of natural environments. A new, committed agency, with many of its leaders tested in the crucible of Hetch Hetchy, thwarted the ambitions of the bureau and the corps.

In the 1919 "Report of the Director of the National Park Service," written by Albright and endorsed by Mather, the agency identified the "menace of irrigation projects." The agency began to define its principles. Water storage was inappropriate in Yellowstone, Glacier, and other parks because of the destruction of natural environments, creating artificial landscapes and a general eyesore. But there was a loftier question at stake regarding the parks and the nation. Albright did not shirk from asking it: "Is there not some place in this great Nation of ours where lakes can be preserved in their natural state; where we and all generations to follow us can enjoy the beauty and charm of mountain waters in the midst of primeval forests?"[17]

Secretary Lane did not receive such challenging questions gracefully. The secretary saw the legacy of Hetch Hetchy as a precedent for invasion of the parks. He was an irrigation man, aided by another San Franciscan of like mind, Assistant Secretary Alexander Vogelsang. Lane subscribed to the idea that man ought to dominate and control nature. Water running unchecked to the sea represented waste. When Lane gave the commencement address at Brown University, he revealed his resource philosophy. "Every tree is a challenge to use," pronounced the secretary, "and every pool of water and every foot of soil. The mountains are our enemies. We must pierce them and make them serve. The sinful rivers we must curb."[18]

The showdown between Secretary Lane and Park Director Stephen Mather came in Yellowstone National Park's Bechler Valley. Although its scenery was not as spectacular as that of Hetch Hetchy, it did share certain characteristics. It was isolated and few visitors had any knowledge of it. It sat in a national park. Most important, it could be dammed to create a reservoir for irrigation water, and in 1919 that was exactly what Idaho sugar beet farmers hoped to do. Idaho representative Addison Smith subscribed to the idea that national parks were open to commercial invasion for a justifiable reason. In leading the farmers' fight, Smith maintained that the proposed Bechler Valley dam would do no noticeable damage to the scenic values of the park. Smith's arguments echoed those of Montana senator Thomas Walsh, who was busy trying to get Congress to authorize a low dam on Yellowstone Lake to please his farming constituents. Both legislators believed that the gains would far outweigh the minor park incursions. In Congressman Smith's view, the Bechler Valley was nothing more than a miserable swamp, inhabited only by swarming mosquitoes and so isolated that nobody ever visited it. It was worthless land, unworthy of protection, since "absolutely nothing in the way of unusual scenery or other interesting features" existed in this corner of the park. Measured against the needs of struggling downstream farmers who had suffered a drought and crop loss, the relinquishment of some 8,000 acres of isolated park swampland seemed insignificant indeed. Whereas the proponents of the Hetch Hetchy dam had argued for the needs of urban women and children, Representative Smith conjured up an equally potent icon—the American farmer and his family.[19]

Secretary Lane backed the Idaho farmers. He had fought hard for the Hetch Hetchy dam, and now he worked in behalf of these Rocky Mountain tillers of the soil. Within the Department of the Interior the lines were drawn. Stephen Mather stiffened his resolve, and both he and Albright coaxed, argued, and invoked "the memory of the 'Hetch Hetchy steal' of 1913."[20] Albright wrote an eight-page memorandum defending the sanctity of Yellowstone, which Lane melodramatically tore in half and threw in the wastebasket.[21] It appeared that Secretary Lane would fire the two leaders of the National Park Service, but quite fortuitously, from the park defenders' view, Lane left his post and Washington, accepting a more lucrative position with the Pan American Oil Company. In his place came John Barton Payne, who soon reversed his predecessor's position.

To simplify a complex battle, the Department of the Interior turned away both irrigation proposals to place dams in Yellowstone National Park, and by 1923 they disappeared. Secretary Lane faced determined opposition in the very persons he had recruited to run the National Park Service. Mather and Albright had risked their positions to uphold the sanctity of Yellowstone Park.

The National Park Service did not stand alone. In the Bechler Valley fight, the National Parks Association, under the leadership of Robert Sterling Yard, developed arguments denied to a federal agency. For instance, Yard suggested that these clever farmers had organized as irrigation districts and intended "to sell national park water," since they would really need it only in drought years. Yard portrayed the farmers not as struggling individual families but as enterprising Mormon communities banding together to capture a valuable federal resource. Capitalizing on a widespread prejudice against Mormons, Yard noted that "strong political influences in Utah will get behind the Fall River Basin Bill at its next appearance in Congress."[22] Like its Hetch Hetchy predecessor, the Society for the Preservation of National Parks, and unlike an official government agency, the new association could engage in vigorous debate. This debate was not always objective, and Yard was not above belittling the interests of farm folks for what he considered the nation's greater need for primitive recreation. In a sense the Yellowstone dam controversy represented the triumph of East and West Coast elites over the rural interests of people who had to make a living from their environment. As with Hetch Hetchy, the issue was a matter of priorities. Were the resources of Yellowstone to be exclusively the domain of tourists, or were the needs of farmers to be respected and the local economy encouraged?[23] In Yellowstone, the tourists prevailed.

Hetch Hetchy remained in the background of the Yellowstone conflict, muted yet significant. When *The Outlook* magazine editor Lyman Abbott first broke the dam story in 1920, he captioned his account "Another Hetch Hetchy." What followed was a 500-word editorial describing the Yellowstone situation, but not one mention of Hetch Hetchy. One can only presume he believed that the Hetch Hetchy fight was so familiar to his readership that explanation was unnecessary and that a mere reference to the famed battle was sufficient to alert the reader to another invasion.[24] When geographer Michael J. Yochim concluded his account of the Yellowstone Park dam battles, he suggested that the fight could not have been won without the example of Hetch Hetchy. The Raker Act, though hardly an easy victory for San Francisco, showed that a national treasure could be lost. "Yosemite provided an illustration of what *not* to do in Yellowstone."[25]

While the National Park Service fought to retain an undeveloped Yellowstone, in Yosemite National Park it sought to minimize the impact of the Hetch Hetchy construction. At issue was whether San Francisco, having won the right to inundate the valley, would extend its hegemony over one-third of Yosemite National Park. Whether disagreement occurred over pleasure boating, trail building, construction cleanup, or something else, the true issue was power. Would San Francisco or the National Park Service control the northern section of Yosemite National Park? The city believed that the conditions of the Raker Act gave it control of the reservoir, especially since it owned the land beneath the water. Furthermore, because of the paramount issue of sanitation and public health, the city argued that it must control the high-country watershed of the Tuolumne River.

Also in question was whether the National Park Service would demand that San Francisco fulfill its obligations under the Raker Act. The act was lengthy and complicated, and as San Francisco fought for passage, it made concessions it would come to regret. These concessions were so onerous that many in the bay city would come to question the wisdom of developing the Hetch Hetchy Valley. Costs mounted, and the National Park Service insisted on compliance with the stipulations of the act. Control was at issue as much as money. How extensive would be the damage to the flora and fauna of Yosemite National Park? Would the public have access to the watershed for recreational purposes? How would the conditions of the Raker Act play out on the ground? Who would have control the Tuolumne River watershed, a land mass of 420,000 acres? City Engineer O'Shaughnessy thought the Raker Act extended a *right* not only to dam the valley but also to control the vast Tuolumne watershed. He believed the 240,000 acres of park land (the other 180,000 acres was in Stanislaus National Forest) should be the city's preserve. The rangers of Yosemite National Park, on the other hand, believed the Raker Act merely granted the city the *privilege* of capturing the waters of the river in a reservoir. The city's jurisdiction ended at the water's edge. In addressing these issues, the National Park Service policy focused on minimizing the damage of the project while retaining public access to the land and water in question. San Francisco fought to maximize its control, excluding the public and making the watershed its private domain.

From San Francisco's perspective, it seemed that Yosemite Park held all the cards. The city would have to conform to all regulations. It would be required to pay for the timber it cut along rights-of-way. No construction or work on park lands could commence until the city filed maps and permit applications and the secretary of the interior approved them. The public had to be given free use of roads constructed by San Francisco. The government would receive free use of the phone lines constructed and maintained by San Francisco. The Raker Act required the city to build an extensive trail system as well as a road circumventing the Hetch Hetchy Reservoir. Five years after passage of the act, the city commenced paying the United States government $15,000 annually for 10 years. That fee increased to $20,000 in the next 10 years, and finally to $30,000 unless adjusted by Congress. San Francisco paid this "rental fee" to Yosemite Park but had no say in how the money would be spent. The city had to provide water for close-by park buildings and campgrounds. It also had to guarantee the Modesto and Turlock irrigation districts abundant water, honoring their senior water rights. The city could not begin to use Hetch Hetchy water until it made full use of the waters of the Spring Valley Water Company, thus protecting the company and forcing the city to finally purchase Spring Valley in 1928. Furthermore, section 6, which prevented sale of Hetch Hetchy power to PG&E, was particularly galling.

Also, San Francisco received no federal help, save the grant to construct the dam. In an era when the Bureau of Reclamation allocated many millions of federal dollars for construction of new dams, San Francisco had to go it alone. Of course, in theory, the farmers and ranchers would repay the Reclamation Service through water fees, but that rarely happened. While communities throughout the West fed at the federal trough, San Francisco was going hungry and actually paying a yearly fee to Yosemite Park for the privilege of water storage. San Francisco supervisors bristled at what they saw as unfairness, especially as they watched Oakland and the participating East Bay cities develop their own water system in the late 1920s with privileges and a streamlined federal permit system, a result of the 1920 Water Power Act.

Of course the provisions of the Raker Act were one thing; the enforcement of those provisions was another. To the city's dismay, both the U.S.

Army (until 1915) and then the National Park Service and the Department of the Interior were determined to protect the park. There was little sympathy for San Francisco. The administrators of Yosemite looked on the city as a grasping antagonist, a piranha bent on eating away the flesh of Yosemite Park for its nourishment. Park Service personnel studied and understood the Raker Act and constantly reminded San Francisco when it evaded its responsibilities.

The city seemed most remiss in its obligation to build roads and trails in order to open the Hetch Hetchy area to recreational tourism. Director Stephen Mather, with his dedication to a usable park, considered City Attorney Percy Long's promise that his city would "build at its own expense a magnificent system of roads and trails which will make one of the most beautiful scenic parts of the Sierras" to be one of the only positive sections of the Raker Act. At first Mather said nothing, allowing Long and his colleagues to postpone such projects given the city's financial difficulties and the tough labor climate created by World War I. But by 1920 the director reminded Long, O'Shaughnessy, and the supervisors of the act's provision that the city would construct roads and trails. San Francisco leaders, struggling with mounting debt and the difficult construction deep in the mountains, responded with surprise and outrage at such an imposition. To assuage the head of the National Park Service, the city conferred the name *Mather* on the upper terminal of the Hetch Hetchy Railroad. However, such a transparent move only made the Park Service chief more determined to see that the city carried out its obligations.

For its part, the city chastised the Park Service for its failure to maintain good roads. In November 1918 O'Shaughnessy complained to Yosemite Park superintendent Washington B. Lewis that the roads in the northern part of the park were in bad shape. O'Shaughnessy believed "that for a payment of a toll by San Francisco of $15,000 a year," the roads ought to be repaired. In a crusty mood, the city engineer stated that the Big Oak Flat Road should be improved "as the people of San Francisco, who are paying the bills, should get some measure of relief and benefit from the money they are contributing to the Park."[26] Of course the fee paid had no stipulation, except that it would be used in Yosemite Park.

When the city completed the dam in May 1923, Yosemite Park superintendent Lewis wrote O'Shaughnessy with a compromise regarding roads. Rather than releasing the dam construction crews, why not redirect their effort to road and trail construction? But the city had lost interest. Months of inactivity became years of frustration for Superintendent Lewis, who grew increasingly suspicious and disillusioned with San Francisco's noncompliance.

Certainly the city dragged its feet because of financial considerations, but it also had a vested interest in keeping the Tuolumne River watershed a de facto wilderness area. The Spring Valley Water Company had successfully cordoned off its extensive lands on the San Francisco Peninsula, and once the city purchased SVWC, it continued a policy of no swimming, boating, or even walking. The area was, and is today, closed to public entry.[27] San Francisco wished to do the same with the vast 420,000-acre Tuolumne River watershed. The area represented an ecological unit, and if the city could gain hegemony, it would guarantee the purity of its water. In 1927 the city posted some of the park watershed area as closed to campers, fishermen, and recreationists. The Park Service rangers removed the signs. The *San Francisco Chronicle* jumped into the jurisdictional dispute by proclaiming that "instead of roads and trails to take people into the Hetch Hetchy watershed there ought to be fences and gates to keep them out."[28] It became evident that the city—if it could get away with it—wished to declare the whole Tuolumne River watershed as its preserve. To ensure pure water, the city municipal utility district would restrict human use whenever possible.[29] In response park officials announced that if hikers, campers, and recreationists polluted the water, San Francisco would have to do what many other cities did—chlorinate it. When Stephen Mather suggested that the city might have to build a filtration plant and chlorinate its water, San Francisco supervisor Keenan, a man of short temper, exploded: "Do you mean to say that the federal government wants San Francisco to build a lot of roads so that people can go to the lake and pollute the water, endangering the lives of 750,000 people just for the benefit of a few thousand?"[30] The answer was yes. Human use of watersheds was common enough. Neither Stephen Mather, Horace Albright, nor Superintendent Washington Lewis was willing to shut down one-third of the park for the convenience of San Francisco. Director Mather became increasingly perturbed by San Francisco supervisors ignorant of the Raker Act. They seemed to come and go with the regular cycle of municipal elections. In 1928, after a difficult session with the supervisors, Mather, in a jocular mood, mentioned to a reporter that it seemed "rather difficult to make them understand that there was such an act and that it imposed upon San Francisco certain obligations."[31]

Although few of the supervisors understood the Raker Act, some of them did enjoy the outdoor benefits of the Hetch Hetchy region. Assistant Director Horace Albright was more than a little upset when he heard that some supervisors and guests were driving the road from O'Shaughnessy Dam to Lake Eleanor, a locked route for emergency use only. Not only were these fishing parties driving on a closed road, but "they carried in intoxicating liquor and the City party usually got 'gloriously drunk'"[32] The idea of San Francisco officials excluding the public but using park land for their own private retreat for fishing and socializing infuriated Albright. Nor did such activities escape the pages of the San Francisco newspapers. The *Bulletin* ran a damaging cartoon showing Hetch Hetchy reservoir fenced off, with a sign "Supervisor's Summer Camp for Junket Trips: Public Keep Out" protecting "Supervisors' Rest."[33]

In contrast, when the Yosemite Park and Curry Company, the primary concessionaire in Yosemite Valley, applied for a permit to begin pleasure boat trips on the Hetch Hetchy reservoir, San Francisco's city attorney Robert Searles voiced official opposition. The lake was "exceedingly dangerous for the operation of pleasure-boats, due to the high winds which prevail almost daily." O'Shaughnessy visited Yosemite Park assistant superintendent Earnest Leavitt and warned him against small boats, stating that "they would surely be upset and the occupants drowned, and our sad job then would be to pick

FIGURE 25. City Engineer Michael O'Shaughnessy tended to think of the Tuolumne watershed as the city's private domain. The National Park Service disagreed. *Courtesy of the SFPUC.*

out the dead bodies which might pollute the water."[34] Albright inquired of Superintendent Lewis if Searles and O'Shaughnessy's scare tactics were valid. In a lengthy letter, Lewis replied that he had "never observed anything in the way of wind and storm conditions which would constitute a danger in the operation of boats." Neither had his rangers—more familiar with the reservoir than he—ever observed such conditions. In the final regulations, shore fishing would be permitted but not boating. However, the boating ban was for reasons other than the exaggerations advanced by San Francisco.[35]

In spite of negative publicity and the wrist slapping by Mather, the city leaders dodged their responsibility, seeking protection from clever city attorneys or their congressional delegation. But there was no avoiding reality, and San Francisco officials faced a blistering attack even in their own newspapers. The *San Francisco News* devoted a full-page editorial titled "Complete Hetch Hetchy or Lose Everything." In a powerful, yet measured, message, the paper attacked leaders, especially the dominating, domineering, "bull-headed" Michael O'Shaughnessy, for alienating Washington and humiliating the city in the halls of Congress.[36]

Representative Lewis Cramton of Michigan administered the humiliation. A capable and devoted friend of the national parks, Cramton was outraged by San Francisco's attitude and actions. He fully understood the Raker Act, for he had been a member of the Congress that enacted it. He was good friends with both Mather and Albright and followed the Hetch Hetchy development with interest and foreboding. He made at least two congressional investigative trips to Yosemite National Park to confer and observe. In 1928 he delivered a savage blow to San Francisco in the halls of Congress, one that would leave the city reeling and practically begging for compromise and reconciliation. Cramton charged that the city was "interfering with the administration of the great Yosemite National Park and trying to stop the legitimate use of it by the public." Furthermore, the city was "holding off and failing to build the roads that were expressly provided for in that (Raker) act, estimated to cost $1,680,000."[37] He also attacked the power sale arrangement between the city and the Pacific Gas and Electric Company. San Francisco had, in a word, broken with the spirit and letter of the Raker Act. At another point, and after a blistering attack, Cramton dropped the threat that "if San Francisco cannot properly cooperate with its generous landlord, the relationship had better cease and the Federal Government resume exclusive use of the park area."[38]

San Francisco's public response to Cramton was understandably acrid. Supervisor Frank Havenner ranted that he would challenge "the right of the United States government to continue recreational activity or permit development in Hetch Hetchy or Lake Eleanor regions, even if the whole public must be barred." Supervisor Alfred Roncovieri declared that San Francisco would move at once to have the Raker Act amended in favor of city control.[39] Actually, San Francisco had few grounds for argument. The Raker Act, while stressing clean water and public health, recognized that the region must be open to recreation. That was, of course, the reason for section 9(P), requiring the city to build roads and trails. Furthermore, the Freeman Report, the pro-San Francisco script that influenced Congress in 1913, stressed the tourist and recreational benefits which would accrue from the reservoir. Roads, hotels, and campgrounds overlooking the reservoir would bring tourists to a spectacular site. In 1928 the supervisors and the city engineer seemed to have conveniently forgotten the very arguments that abetted the city in gaining the grant. Closing off Hetch Hetchy and the Tuolumne River watershed was never the intent of the Raker Act, and both the National Park Service and the California public reminded San Francisco of that fact.

The city could muster little support in California and surrounding communities. Many residents of the hinterland towns distrusted the grasping city. Although San Francisco cultivated good relations with the San Joaquin Valley irrigation interests, many towns closer to the Hetch Hetchy project harbored deep resentments. The *Sonora Banner* labeled San Francisco's actions a "grand theft of public property" claimed that and if they succeeded, all of Stanislaus National Park [Forest] could be closed to campers and tourists. San Francisco might have the advantage through "bullyrag, and bamboozle and bribe" of local interests, but the people of the region would not give up without a fight.[40]

The *Banner* editors need not have been so fearful. The determination of the National Park Service, the outspoken revelations of Representative Cramton, the aggressiveness of O'Shaughnessy, and the palpable ignorance of some of the San Francisco supervisors all worked against the city. It retreated. By 1929 O'Shaughnessy directed his efforts to finishing the Hetch Hetchy system rather than wasting energy in jurisdictional squabbles with Yosemite Park superintendent Washington Lewis or his successor, Charles G. Thomson. Most encouraging, San Francisco leaders, including Lewis Byington for the Public Utilities District, O'Shaughnessy, and City Attorney John J. O'Toole, hammered out a roads and trails agreement with the secretary of the interior in December 1930.

Under this agreement the city would complete a system of trails on the north side of Hetch Hetchy Reservoir, making the high country once again accessible to hikers. The city reconstructed the road from Mather (the former Hog Ranch) to the O'Shaughnessy Dam, providing a good auto road for tourists at an estimated cost of $200,000. The bulk of the monies for roads, however, would be used at the discretion of Yosemite National Park The city pledged six payments of $250,000. Once completed, the city would be free of any obligations. The Park Service soon abandoned Freeman's impractical plan to construct a road around Hetch Hetchy Reservoir, diverting the payments to rebuilding the western and eastern sections of the Tioga Pass road.[41]

Most significant, the agreement stated that "the Department of the Interior alone will be responsible for the maintenance of law and order, regulation of recreation, preservation of water shed, sanitary control, and all other matters of administration of Yosemite National Park."[42] From 1930 on, there would be few disputes regarding jurisdiction. The Raker Act surely gave away an irreplaceable wonder of nature, but the many restrictions assured that *if* the Park Service persevered, the legacy of loss would be circumscribed. The vast wilderness that comprises the northern part of Yosemite National Park remained open to people and provided habitat for wildlife. The agreement confirmed a new precedent—which should need no confirming—that no private or public institution, even though it had acquired certain rights, would be allowed to dictate its will on a national park. Such a precedent, of course, would have to be continually defended and reaffirmed.

HETCH HETCHY'S LEGACY is surely political, but it also has an intellectual side. The controversy established important precedents regarding national policy not only in dam building but also in the origins of land use ideas. However, there are two ideas that are associated with the Hetch Hetchy controversy but should not be. The most important of these fictions is that the Hetch Hetchy fight pitted the forces of development against those favoring wilderness preservation. Roderick Nash, in his classic book *Wilderness and the American Mind* (1967) frames the Hetch Hetchy controversy as one that put the "principle of preservation of wilderness" to the test. In a memorable sentence, Nash defined the issue: "For the first time in the American experience the competing claims of wilderness and civilization to a specific area received a thorough hearing before a national audience."[43] Other historians have followed Nash's lead, one suggesting that Hetch Hetchy was essentially about "salvaging islands of wilderness." Another maintained that Hetch Hetchy's "claim to distinction was wildness," while yet another believed that the defenders of Hetch Hetchy were "champions of the right of all Americans, living and unborn, to experience the vanishing wilderness." Recently, a prominent environmental historian wrote that "Hetch Hetchy became the battle cry of an emerging movement to preserve wilderness."[44] The documentary evidence suggests otherwise.

Of course, *wilderness* is a slippery concept. Native American peoples have noted that the term is ethnocentric. What was wilderness to a European American was home to a Native American. Moreover, the word has changing meanings, and definitions abound. Roderick Nash points out that "one man's wilderness may be another's roadside picnic ground."[45] Some might see wilderness as their suburban backyard or a tent on top of an apartment building. I once had a colleague who defined a wilderness experience as the walk between his air-conditioned house and his air-conditioned automobile. On a more serious note, many within the academic community see wilderness as a culturally constructed notion that is at best romantic, at worst racist, exclusionist, and ahistorical. However, in the context of the Hetch Hetchy historiography, the Wilderness Act of 1964 can provide a workable definition. To be eligible for the American wilderness system, an area should be at least 5,000 acres in size, be "effected primarily by the forces of nature," and include "outstanding opportunities for solitude or a primitive and unconfined type of recreation." It is to be "an area where the earth and its community of life are untrammeled by man, where man himself is a visitor who does not remain."[46] This definition assumes that wilderness is a place, not an idea.[47]

Yet, would Colby and the defenders agree with this 1964 definition? For instance, would a two-track wagon road into Hetch Hetchy Valley have eliminated the "wilderness"? Would a few simple cabins have disqualified the valley from wilderness status? It is impossible to define precisely what the defenders thought wilderness was, but perhaps we can suggest what it was *not*. For Muir his beloved Yosemite Valley was not wilderness. As early as the 1870s the roads, carriages, hotels, and many creature comforts in the valley eliminated it from such consideration. To experience wilderness, Muir escaped the valley and what he considered the annoying, lazy, superficial tourists. By 1912 much more development had taken place, and when automobile drivers requested entrance into the valley, neither the Sierra Club nor Muir opposed the idea, knowing that any wilderness characteristics of Yosemite Valley had already been compromised in the name of tourism.

More to the point, the Hetch Hetchy defenders did not use the word *wilderness*. Preservation of wild lands, however slippery that concept, was not among their objectives. We can search the Colby Brief, hundreds of letters, and the 1909 tract and not find the word *wilderness* anywhere. Most congressional representatives and senators avoided the designation. When they explained their preservationist positions, they referred to "playgrounds," scenery, scenic values, places of "solitary loneliness," "beauties of nature," "picturesque spots," and "beautiful natural features," but seldom wilderness. The use of such descriptive words suggests that the effort to save the valley had economic, commodity implications. The defenders laid their arguments on the rock of tourism.

The defenders of Hetch Hetchy Valley first wanted to save the valley from the dam. If they succeeded, specific plans for its use would come later. Yet in 1907 the Sierra Club officially endorsed road construction to the valley. In 1909 they restated this commitment, noting that Yosemite could be a tourist "gold mine" and Hetch Hetchy was one of "its most priceless attractions." Moreover, the nature lovers suggested a road tour of Yosemite National Park that would seem a sacrilege today. Historian Marguerite Shaffer defines *tour,* from the Latin *tornis,* as a "circular journey." The Sierra Club advocated construction of a circular road, somewhat like that of Yellowstone, beginning at Yosemite Valley, going up the Merced River and over Tuolumne Pass, and dropping into the meadows. From Tuolumne Meadows this scenic drive would descend through Tuolumne Canyon to the Hetch Hetchy Valley, creating an unexcelled mountain drive connecting the three most notable scenic features of the park: Yosemite Valley, Tuolumne Meadows, and the Hetch Hetchy Valley. The Colby Brief advocated a road from Hetch Hetchy to Tuolumne Meadows. In 1908 Marion Randall Parsons suggested a "possible cañon highway from Yosemite to Hetch Hetchy." Two years later Sierra Club member Lucy Washburn enjoyed a horseback trip from Yosemite Valley to the Tuolumne watershed but then suggested that "when our Government shall have built a road over the pass, as is perfectly feasible, and some hostelry shall perch among its beetling snowy crags, the thousands who now see only the Yosemite Valley below it will never fail to see this wild Alpine glory."[48] The club, in short, favored development in the hope that more Americans might come to the valley for recreation and inspiration. Joseph L. Sax observed that "the preservationist is not an elitist who wants to exclude others . . . he is a moralist who wants to convert them."[49] Certainly this was true of Muir, who saw the mountains as a leisure resource and a place of conversion. We forget that the original mission statement of the Sierra Club (1892) was to "explore, enjoy and *render accessible the mountain regions of the Pacific Coast.*"[50] [Italics mine]

The strongest evidence of the valley defenders intent to develop the valley may be found in the Colby Brief, written by William Colby as a critique of the Freeman Report.[51] Colby argued that Yosemite, and by inference all national parks, should be inviolate—indeed, should be sacred spaces. However, he did not equate wilderness with sacred spaces. He saw no contradiction between special places of nature and tourism. Colby contended that the future value of tourism would outstrip that of water. In 50 or 100 years "the need of the Nation for Hetch Hetchy Valley and the extensive camp and hotel sites on its floor will be greater than the need of San Francisco for its use as a reservoir site." He envisioned a developed valley, complete with hotels, restaurants, campground, trails, horses, and fishing opportunities, all within the lush valley and the ever present cliffs and waterfalls. To emphasize the potential commodity value, Colby looked to Switzerland, where the Alps were "practically unfrequented at the commencement of the 19th Century and now its visitors are counted by millions."[52]

Clearly, the accepted tourist model was Switzerland. Scenic mountain regions in the Sierra Nevada and the Rocky Mountains were often labeled the "Switzerland of America." When religion professor William Badè researched his biblical interests in Europe, he took time to study tourism in Switzerland. When he returned to the United States to testify in 1910, congressmen listened intently because he argued the tourist-dollar value of Hetch Hetchy. Switzerland, long considered to contain the epitome of mountain scenery, was the benchmark by which tourists judged American landscapes. The Alps, while spectacular, had no waterfalls comparable to those of Yosemite or Hetch Hetchy valleys. The Matterhorn met its match with the Grand Teton range of Wyoming. Allen Chamberlain, in arguing for campgrounds, hotels, and conveniences in the Hetch Hetchy Valley, reminded his readers that "Switzerland regards its scenery as a money-producing asset to the extent of some two hundred million dollars." The message was clear: Although the American West held scenery comparable or superior to that of the Swiss Alps, tourism brought in a fraction of the Swiss take. At the very time Californians should be developing and advertising the Hetch Hetchy Valley as a tourist attraction, San Francisco had bamboozled the state into its water agenda.[53]

If Hetch Hetchy Valley had escaped the fight unscathed, what would it look like today? If we could look down on the valley from LeConte Point, in all likelihood we would see a near clone of Yosemite Valley. Perhaps there would be more car camping and fewer luxury hotels, such as the Ahwahnee. Yet there would be motels, eateries, and a considerable infrastructure to service the traveling public. Traffic jams might not exist, but automobiles would be very much in evidence, allowing visitors to approach scenic areas with a minimum of walking.

Of course we must consider the possibility that the Colby Brief was an insincere ploy to mask the club's true wilderness agenda—a subterfuge to curry support by emphasizing development and supporting commercial tourism.[54] Perhaps Colby, Chamberlain, Muir, and others had to make repugnant compromises, with the thought that they could be reversed at a later date. However, in letters and strategy sessions, there is no evidence that the Sierra Club leadership attempted to hoodwink the public or Secretary Walter Fisher. The acceptance of roads and hotels was sincere—a realization of the changing nature of tourism and commodity conservation, not a planned strategy for the exigencies of the moment.

In fact, no one within the club or its friends argued against the Colby Brief. Bostonian Allen Chamberlain, who visited Hetch Hetchy in 1909, had written in a 1910 issue of *The Outlook* that he favored extension of the present road nine miles and construction of hotels or at least tent cabins to open the valley. Chamberlain not only advocated opening Hetch Hetchy to tourism but also believed that all of the national parks must be made into "attractive vacation resorts." He wanted the American public to enjoy nature, but he also wanted to save the parks from commercialism. Unless the public became more aware of the parks, "selfish interests are likely to steal an important part of our birthright."[55] Chamberlain understood, as did the preservationists, that the public needed mountains, but the mountains also needed the public. To enjoy mountain scenery, the public demanded roads. When Muir pronounced his famous dictum that "thousands of tired, nerve-shaken, over-civilized people are beginning to find out that going to the mountains is going home; that wilderness is a necessity," he freely admitted that they would get to the mountains by "means of good roads."[56] We can only conclude that the defenders of Hetch Hetchy wished its scenery opened to mass tourism, using the Swiss model of access through technology.

If the Sierra Club position for development was clear, why have both scholars and the public believed that this was a wilderness controversy? Without question, the passionate leadership of John Muir would suggest this interpretation. No American has been more associated with wilderness than Muir. When a young Muir dropped into the Hetch Hetchy Valley in 1871, he saw it as pure wilderness, and he described it as such, unaware of the cultural impact of the Miwok Indians. He relished the wilderness experience with artist and friend William Keith in 1895 and again in 1907. On an earlier occasion, he approached Hetch Hetchy by hiking down the Tuolumne Canyon. He described the rollicking river as well as the formidable boulder-filled gorge, with its horrendously difficult descent through a canyon wilderness that more than hinted at danger. These experiences, and the way he wrote of them, could easily convince the reader that Muir not only experienced wilderness but wished to retain it, as well. Muir was a prophet of wilderness, an evangelist who used the word at almost every opportunity.

And yet he read the Colby Brief and did not challenge or change a word of it. In some respects the legendary Muir, the mythic Muir, was so clearly the spiritual leader of the Hetch Hetchy fight that the more practical, down-in-the-trenches William Colby has often been overlooked. So have the intentions of the valley defenders who accepted the idea of opening the valley to tourism and creating a mirror image of Yosemite Valley, in both its features and its concessions to boosterism, tourism, consumerism, and the domestication of nature. It was not that the valley defenders did not believe in the attributes of wildness, but that the valley was not the right place and 1913 was not the right time.

Because of Muir's warm embrace of a biocentric view of nature and his writings on "Lord Man," as he derisively referred to the arrogance of human beings, it is easy to believe that Muir and the defenders of Hetch Hetchy thought they were fighting for the rights of nature. This, I freely admit, is a myth of my own making. When I first began research, I was convinced that material abounded suggesting that Muir's campaign preceded Aldo Leopold's call for a new "land ethic," extending rights to the flora and fauna of the Hetch Hetchy Valley. However, none of the pamphlets, speeches, or testimony suggests that this was anything more than a fight for the rights of humans. Much as Muir believed in rights for nature and in a certain egalitarianism among all living creatures, I could find no evidence that he ever used such an argument for preserving the valley. In 1912 the American public was not prepared for such an expansion of ethics. Animal life and plant life were entitled to appreciation, but not ethical considerations or legal rights. As early as 1875 Muir wrote a friend that "to obtain a hearing on behalf of nature from any other standpoint than that of human use is almost impossible."[57] To argue for preservation of Hetch Hetchy on the basis of the rights of nature would challenge the accepted theological view that God created human beings to have dominion over all life on earth. Muir understood this, and even though he had often railed against the presumptions of "Lord Man," he was mindful that a biocentric view might alienate many traditional supporters. It was much safer to blame the greedy capitalists for the undoing of nature. The basis for saving the Hetch Hetchy Valley would not be ecological, although certainly there is a strain of stewardship in their arguments. At the philosophical center the rationale for protection from bulldozers and steam shovels would be the valley's usefulness to human beings in spiritual ways. Roderick Nash noted that when it came to politics and Hetch Hetchy, Muir "tempered his biocentricity" and "camouflaged his radical egalitarianism in more acceptable rhetoric centered on the benefits of nature for people."[58]

Thus both Muir and the Colby Brief discounted any biocentric or "deep ecology" view. Muir had been tarred by his enemies as a person devoid of compassion for humans. Mayor James Phelan testified before Congress that Muir would sacrifice his own family for the preservation of beauty. Presumably, nature affected him more emotionally than the thirsty children of San Francisco. To espouse his philosophy with Hetch Hetchy in mind would be self-defeating. However, one can argue that through Muir's eloquent writings and Colby's sincere but more practical efforts, there is an *implied* right of nature. Muir's chapter on Hetch Hetchy Valley in *The Mountains of California* hints at the possibility that to destroy the valley would be a crime, indeed a sin, against nature. Muir proclaimed his passion of attachment, not to a supernatural world, but rather to a natural one: "Dam Hetch Hetchy! As well dam for water-tanks the people's cathedrals and churches, for no holier temple has ever been consecrated by the heart of man." In this impassioned peroration, Muir, by a stunning theological comparison, conveys on the valley an ethical right to exist.[59]

NO ONE APPRECIATED Muir's eloquent words more than David Brower, and no one has taken the Hetch Hetchy legacy more to heart. Brower, perhaps the preeminent environmentalist of the second half of the twentieth century, carried on his river-dam battles with the memory of Hetch Hetchy close at hand. His first full-fledged fight occurred at Echo Park in the early 1950s. At the confluence of the Green and Yampa rivers, within Dinosaur National Monument, the Bureau of Reclamation determined to construct a high dam. The agency prepared to exercise a right guaranteed in 1938 with the establishment of the monument. With local support and the firm belief that every dam, private or public, encouraged prosperity, the bureau was unprepared for the fire of opposition that emerged from small environmental groups, kindled by concern for the national parks and monuments and fueled by the memory of Hetch Hetchy.[60] The parallels between Echo Park and the earlier battle were quite obvious, especially to Sierra Club leaders who determined to strike a lethal blow against the dam plans. David Brower emerged as the leader, a man steeped in the legacy of Hetch Hetchy. Brower never knew John Muir, but in many ways he was a carbon copy in his spirited defense of nature and the parks. Muir was a model not only as one who had spent his life in the pure enjoyment of the mountains but also as an elegant defender of nature. Brower was cut from the same cloth. He felt deeply the loss of Hetch Hetchy, and "he seemed to hear echoes of Muir resounding in the canyons of the Green and Yampa rivers, calling him and the Club to act."[61] They acted. Summoning all the energy the little club had, Brower led them in a campaign meant, in a sense, to make amends for the wrong of Hetch Hetchy. He would see that it did not happen again.

But Hetch Hetchy did not haunt David Brower alone. Harold Bradley had been a young man in 1913, but he remembered the fight well and even wrote a scolding letter to Norman Hapgood, editor of *Collier's Weekly,* for supporting San Francisco's position. Now, almost 40 years later, for six wonderful days he floated the Yampa River with his sons and, perhaps even more important, a movie camera. His film of a family adventure running the rapids, admiring the immense sandstone cliffs, and camping on the pristine sandy beaches—all destroyed and submerged under water if the Bureau of Reclamation built the dam—brought to life the beauty of a country almost as isolated as Hetch Hetchy in 1910. Bradley, whose father worked closely with Muir, must have felt a paternal presence as he made hundreds of friends for the dinosaur country through his film. The two fights were so similar, he reflected, "that the campaign literature on both sides might be interchanged, with the appropriate names added."[62]

As the battle over the Echo Park dam proposal wore on, Brower became more and more determined that the error of Hetch Hetchy would not be repeated. To a Senate subcommittee, he testified that "if we heed the lesson learned from the tragedy of the misplaced dam in Hetch Hetchy, we can prevent a far more disastrous stumble in Dinosaur National Monument." But Brower knew, perhaps more than Muir, the power of images. The *Sierra Club Bulletin* soon featured photographs of Steamboat Rock, the Tiger Wall, and other features of the Echo Park area. Perhaps his most potent propaganda piece was the film *Two Yosemites.* The film started with beautiful footage of Yosemite Valley. It then revealed "Little Joe" LeConte's still photographs of Hetch Hetchy Valley before it was stripped of vegetation in 1917, juxtaposed in stark contrast with Philip Hyde's photographs showing an inundated Hetch Hetchy Valley at low water. The late incarnation of the valley thus appeared as mudflats, tree stumps, and dust devils swirling about what Muir would call "the graveyard of Hetch Hetchy." No one could miss the message. Wilderness Society lobbyist Howard Zahniser carried the film with him as he circulated the halls of Congress, setting up his projector for any legislator who would take 11 minutes to watch it. Another use of the Hetch Hetchy heritage came with the publication of Robert Cutter's *Hetch Hetchy—Once Is Too Often.* The slim book reached the 9,000 members of the Sierra Club as well as at least another 1,000 through special mailings.[63]

Echo Park did not become another Hetch Hetchy. By 1956 the Bureau of Reclamation abandoned its plans. Today one can visit Echo Park by rough road or, preferably, by raft either from the Lodore Canyon of the Green

River or the Yampa River. A determined coalition of conservation organizations, undoubtedly motivated by the memory of Hetch Hetchy, fought and won. In the Echo Park controversy the sublime Sierra Nevada valley took on a new meaning. It was lost, yet in a way, it lived. A despondent John Muir had predicted that some good would come from the loss of Hetch Hetchy, and indeed it did.

The Echo Park struggle was as much the beginning as the end of dam controversies on the Colorado River. Well below the confluence of the Green River and Colorado River lay Glen Canyon. Spectacular but unknown, Glen Canyon was unprotected by any national park system designation. In their efforts to save Echo Park, David Brower and the environmentalists tacitly consented to a compromise in which Glen Canyon would be sacrificed to save Echo Park. Brower deeply regretted that compromise for the rest of his life. It was, he remarked with remorse, a "flagrant betrayal, unequaled in the conservation history that sixty-eight years of *Sierra Club Bulletins* have recorded." Brower blamed himself, but realized the difficulty when faced with a huge bureaucracy that could manufacture plausible statistics and "hold press interviews faster than a true interpretation can overtake them."[64] The defenders of Hetch Hetchy who endured the San Francisco publicity machine would nod their heads in agreement.

The Southwest's water and power seekers were not satisfied with the Glen Canyon Dam and Lake Powell. A growing population, particularly in Arizona and Southern California, placed pressure on politicians, who in turn encouraged a willing Bureau of Reclamation. In the 1960s the ghost of Hetch Hetchy once again emerged in the form of bureau plans for the Marble Canyon dam and the Bridge Canyon dam. Neither Marble Canyon, just to the north of Grand Canyon National Park, nor Bridge Canyon, south of the national park, were protected by special status. However, both sites were strongly *associated* with the national park and were, geographically, part of the greater Grand Canyon. Again the specter of Hetch Hetchy emerged, and Brower prepared for battle. Sickened by the loss of Glen Canyon, he now had a chance at redemption by saving the Grand Canyon from the scourge of two more dams. On the other side was Secretary of the Interior Stewart Udall, a politician much beholden to his home state of Arizona. However, Udall was environmentally sensitive and had even noted the Hetch Hetchy fight in his popular environmental history, *The Quiet Crisis.* Hetch Hetchy Valley "was flooded out," wrote Udall, "but those who had fought a losing fight for the principles of park preservation served notice on the country that its outdoor temples would be defended with blood and bone."[65] Of course, the lessons of history, even when one writes them, do not always influence one's action. Udall remained committed to supporting the determination of his Bureau of Reclamation director, Floyd Dominy, to raise the two high dams.

In the end, Udall and Dominy could not overcome those who fought with "blood and bone." David Brower organized and publicized his way to victory with the memory of Hetch Hetchy and Glen Canyon as inspirations. However, this time he had powerful allies. Chief among them was Senator Henry "Scoop" Jackson of Washington State, who had distrusted Udall ever since he promoted a plan to import water from the Pacific Northwest to the Southwest. Jackson feared a "water steal" as strongly as Brower dreaded another Hetch Hetchy. In the final analysis, Udall, Senator Barry Goldwater, and the irascible Floyd Dominy could not overcome the opposition. In 1968 President Lyndon Johnson signed the Central Arizona Project bill without the two controversial dams. In future years both sites became protected through the expansion of National Park Service hegemony up and down the great canyon.[66]

LESS TANGIBLE BUT no less significant, there has been a growing appreciation by Americans for national parks and grand scenery. Americans realized that nature, while bountiful, was also finite and, once lost, could not be easily replaced. For some, Hetch Hetchy was a painful reminder. Late in life, Harriet Monroe had not forgotten Hetch Hetchy: "I cannot think of the lake which today covers the flowery vale without a pang for the nation's loss of something too beautiful to perish, one of those miracles which Nature flings out for perpetual delight, and which man too often destroys."[67]

In a sense, Hetch Hetchy was a national awakening. Through organization, through support, through a rather major transformation in societal thinking toward the out-of-doors, no longer could the Bureau of Reclamation or the Army Corps of Engineers expect that their efforts to marry technology with scenic land would go unchallenged.

Harold Bradley remarked that the arguments regarding the Echo Park and Hetch Hetchy dam projects were so similar that only the names need be changed. Not only the arguments but also the modern techniques for convincing the public and politicians owe much to the defenders of Hetch Hetchy. The fight has lost little of its symbolic importance in almost 100 years. It constantly reminds many Americans that the sanctity of our national parks and scenic lands cannot necessarily be taken for granted. There will always be those who can find sufficient reason for invasion of a park or monument. Yet, because of the memory of the Hetch Hetchy fight, such inroads will not be easy, and in most cases, will be impossible.

CHAPTER 10

Restoration

"Waiting in Yosemite National Park, under water, is a potential masterpiece of restoration."

KEN BROWER

"For the restorationist, nature is of intrinsic worth as well as being valuable for human use."

FREDERICK TURNER

For over 50 years environmentalists lamented the loss of the Hetch Hetchy Valley. But what could be done? It seemed a fait accompli, an irreparable mistake. Now there are new questions and new challenges to the dam. Can restoration be the next phase in the evolution of the Hetch Hetchy Valley? Can the great O'Shaughnessy Dam be dismantled? Can the valley come full circle to what it once had been? In 1913 Senator William Borah of Idaho did not think so. He cautioned his fellow legislators that "we ought to pause and consider well our act. What we undo in the way of defacing and marring these marvelous scenes, nature in all her majesty and strength can not restore."[1] Never did Borah imagine that humans could resurrect a canyon once transformed into a reservoir.

However, that was 1913. In an age when construction of dams was just beginning, no one could conceive of removing a dam. Even in 1955, when David Brower made the film *Two Yosemites,* he did not think anything could be done about the O'Shaughnessy Dam. He believed it destroyed Hetch Hetchy for all time. "So," he recently reflected, "I just gave up." Forty-five years later he changed his mind. By then restoration was not only desirable but feasible.[2] When I recently walked across the massive O'Shaughnessy Dam, the idea of breaching and removing it seemed a herculean task. Michael O'Shaughnessy built a dam to last, and with a 298-foot base, about the size of a football field, neither nature nor man could remove it easily. The sheer technological difficulty of breaching the dam, though, has not eased San Francisco's concerns for its security. During World War II the Park Service closed the area to visitors, and six full-time guards patrolled against sabotage. More recently, after September 11, 2001, the city increased security once again, taking extra measures to protect the dam as well as the water supply. However, not all potential foes of the O'Shaughnessy Dam come from overseas. In the last 40 years environmental-radical groups such as Earth First! have protested development through acts of sabotage, such as spiking trees, removing fences, and dismantling construction equipment.

FIGURE 26. In July of 1987 a protester under cover of night rappelled down the side of the O'Shaughnessy Dam, spreading a plastic crack and evoking John Muir's thoughts. San Francisco employees quickly sandblasted the offensive inscription, but not before photographer David Cross recorded the event. *Courtesy of Restore Hetch Hetchy*.

Major dams, such as Glen Canyon, have been the object of political and environmental pranks. O'Shaughnessy Dam has not been spared. In July of 1987, during the dark of night, a climber rappelled down its sheer cement wall, rolled out a long black plastic crack as he went, and then finished off his handiwork with the words "FREE THE RIVERS—J. MUIR." City workers soon sandblasted the graffiti, but not before photographer David Cross recorded the event. Although Muir did not embrace such direct action or civil disobedience, one wonders whether he would have approved of such symbolic action.

This graffiti expressed the frustration of a growing number of Americans who believed that rivers ought to run free and that growth should have limits. By 1960 the new environmentalists looked at dams with a more critical eye. Wallace Stegner, so instrumental in fighting the Echo Park dam proposal, wrote in *Saturday Review* that the structures' disadvantages often outweighed their benefits. He used as a primary example Hetch Hetchy, including a photograph of the O'Shaughnessy Dam with the caption "built over conservationists' protests."[3] Stegner wrote with a purpose. He wanted to undercut the power and authority of the two great dam-building federal agencies, the Bureau of Reclamation and the Army Corps of Engineers. He wished to question projects "begun in narrow engineering terms, or devised by politicians and handed to one or the other bureau for feasibility studies." At that moment Stegner was busy chastising the Bureau of Reclamation and Secretary of the Interior Stewart Udall for advocating the Marble Canyon and Bridge Canyon dams, both of which, if built, would inundate and impact parts of the Grand Canyon of the Colorado River. Stegner wanted a more reflective process—in essence, an environmental impact report that would reflect losses as well as gains.[4] The renowned novelist and historian of the American West ended his message with a paraphrase of Robert Frost's "Mending Wall":

> Before I built a dam I'd ask to
> know
> What I was damming up or drown-
> ing out,
> And to whom I was like to give
> offense . . .

Stegner never suggested, however, that dams ought to be torn down (as Frost suggested concerning his neighbor's fence). Neither the Stanford professor nor any other reputable environmentalist was ready for that step. Rather they adopted a defensive strategy, reacting to feasibility studies with argument and options. Restoration—or in a cruder sense, tearing down a dam—would be proactive. It represented deconstruction, rather than non-construction. And while novelists such as Edward Abbey, in his book *The Monkey Wrench Gang,* fantasied sabotage to remove Glen Canyon Dam, no individual or organization suggested the same fate for Hetch Hetchy throughout the turbulent 1960s.

In 1977, however, a San Francisco journalist-writer named Christopher Swan and his coauthor, Chet Roaman, wrote a book titled *YV88,* an abbreviation for "Yosemite Valley, 1988."[5] Published by the Sierra Club, the fictional account transformed Yosemite National Park into a more environmentally friendly, more primitive place. In the book a light railroad system replaces roads. An expanded El Portal village outside the valley allows for removal of buildings from the park. But most germane, the last section of *YV88* advocates the removal of the O'Shaughnessy Dam and the restoration of Hetch Hetchy Valley. Visitors board the restored Hetch Hetchy Railroad to the newly reopened valley. Swan and Roaman note that increased Hetch Hetchy visitation offsets any loss of electricity income, and Bay Area recycling efforts mitigate the loss of water. While provocative and imaginative, *YV88* was, in the end, fiction. National Park personnel certainly read and discussed the book, but it quickly fell out of favor as too impractical. Certainly among a select audience, the book planted a seed, but much like Ernest Callenbach's *Ecotopia*, it provoked thought among a few, but no action.

A restoration suggestion by visionaries is one thing. The same suggestion advanced by the sitting secretary of the interior is quite another. During the first week of August 1987, Secretary Donald Hodel sat down at his personal computer and hammered out a memo to his staff. He proposed creating "a second Yosemite Valley" through the removal of the O'Shaughnessy Dam and the restoration of the Hetch Hetchy Valley. Two days later he wrote San Francisco mayor Dianne Feinstein, reassuring her that this was just a single step in what could be "a thousand mile journey." Mayor Feinstein did not react warmly to this first step. Publicly she announced that Hodel's idea was "the worst thing since the sale of arms to the ayatollah [Khomeini]." "The idea belongs in Ollie North's shredder." Other prominent Californians immediately weighed in, but on Hodel's side. David Brower called on the mayor and the secretary to "correct the biggest environmental mistake ever committed against the National Park System." William Penn Mott, the high-profile former California park director, was "elated." Yosemite superintendent Jack Morehead thought the proposal "a very exciting idea."[6]

But beyond the initial reaction, officials on both sides were scratching their heads. What did this mean? Why stir up a hornet's nest? Had Hodel experienced some sort of environmental epiphany? Theories ran rampant as to why this interior secretary proposed such an idea. From the perspective of environmentalists, Hodel represented an improvement over the departed Secretary James Watt, but he opposed many environmental programs. Moreover, he was Ronald Reagan's choice, and the former California governor was not known for sympathy to environmentalists' concerns. Probably most would agree with Robert Hackamack, a Sierra Club member of the Tuolumne River group: "Hodel isn't going to propose a deal like this without requiring a compromise that I wouldn't want to compromise on. But it's a wonderful, wonderful thing to talk about."[7]

However, Secretary Hodel insisted that he was serious about the idea. "I have no ulterior motives," he insisted, and claimed that when the first phase of knee-jerk reactions and one-liners dissipated, he hoped to meet "with those people who would like to get serious about talking about it."[8] Meanwhile in Washington, Deputy Assistant Secretary Wayne Merchant tried to satisfy a curious press, lamely explaining that this was only a concept, devoid of any plan. Immediately, however, speculation emerged that the suggestion of restoration of Hetch Hetchy Valley was, as Robert Hackamack suspected, merely a diversionary bone thrown to environmentalists. At issue was the proposed Auburn dam on the American River, a Bureau of Reclamation project. Congress authorized the dam in 1965 at a cost of $465 million. The bureau had spent approximately $330 million in planning and initial construction, but in 1975 an earthquake rattled the Sierra Nevada. Opponents of the dam raised significant questions regarding cost-benefit ratios, but the earthquake sealed the dam's fate, especially since Sacramento would be in the water's path should it fail. But with such an investment, critics suspected that Hodel wanted to resurrect the Auburn dam project. Those given to conspiracy theories believed that Hodel's Hetch Hetchy idea was merely a ruse by the Reagan administration to threaten San Francisco with loss of water while assuring the city that replacement water would come from the Auburn dam—if the city supported the Bureau of Reclamation plan.[9] The *Sacramento Bee* willingly accepted such an interpretation, asking in an editorial, "Is Interior Secretary Donald Hodel's plan for tearing down San Francisco's Hetch Hetchy water system really just another desperate ploy to resuscitate the Bureau of Reclamation's faltering plans for the Auburn dam site?"[10] The *Bee* believed the answer was yes. However, representatives of both Friends of the River and the Sierra Club vigorously denied that the removal was coupled to the Auburn dam proposal. They considered that replacement water could be found from the Tuolumne River by "reoperation" of the river flows. Hetch Hetchy manager Dean Coffey admitted candidly that through "reoperating" and reconstructing the Tuolumne system, the dam could be removed and yet provide for the city's water needs.[11]

Charles Washburn, a professor of mechanical engineering at California State University, Sacramento, supported this view. In his article "No Hetch Hetchy or Auburn Dam," Washburn argued that the first reservoir on any river "yields a large return in water, but as more and larger reservoirs are built, the incremental gains in water yield diminish." The enlarged Don Pedro Reservoir, along with other reservoirs in the Tuolumne watershed, "make[s] Hetch Hetchy redundant as a water conservation reservoir." The Hetch Hetchy Reservoir in 1987 controlled only 26 percent of the river's flow. Washburn believed that "there seems no reason in principle why San Francisco couldn't divert its water from Lake Don Pedro," since the aqueduct tubes run under the reservoir and the gravity system could be retained. Admittedly, the wild and scenic stretch of the river would be changed, "maybe for the worse," but "tearing down the dam might restore good trout fishing."[12]

HAVING DROPPED HIS bombshell on a nervous city, Secretary Hodel disappeared for two weeks in the wilds of Alaska, after vigorously denying that the Auburn Dam influenced his Hetch Hetchy announcement.[13] In Alaska, he unexpectedly encountered a delegation of Sierra Club members at the tiny Deadhorse Airport. A spontaneous, informal meeting resulted. Michael Fischer, the Sierra Club's executive director, took notes:

> He is committed to the principal [*sic*] involved—that of restoration. He sees this as a 10-year project, and is surprised by Dianne's [Feinstein] inability to look beyond the immediate problems raised by the suggestion, especially since she's leaving office shortly. Even if no action on Hetchy is possible in the near future, he agreed to push the idea for its own merit. When Doug [Horn] suggested that David Brower would want to extend the precedent to Glen Canyon Dam, his response was "that day may come."

Hodel expressed his eagerness to work with the Sierra Club and asked that his staff be given access to the club's archives to develop a case for the restoration of the valley. It was a rather remarkable meeting, and Doug Scott paid the secretary a compliment by stating that "only Nixon could have gone to China, and only Hodel could credibly open such a serious discussion." Later, typing up his notes at Sadlerochit Springs, Fischer noted that a caribou grazed within a 100 yards—approvingly.[14]

The analogy to Nixon opening China was apt. It was surprising that Hodel should suggest restoration of the Hetch Hetchy Valley. However, he offered an explanation. During that summer of 1987 the secretary visited Rocky Mountain National Park in Colorado, observing three potentially majestic valleys submerged by small dams. The dams had been constructed by the nearby city of Longmont, but the city faced expensive repairs. In fact two dams might burst. Hodel suggested that they be dismantled by human intervention rather than nature. Perhaps the city of Longmont would agree to their removal if the Park Service purchased the sites and also found replacement water? Longmont agreed. Having participated in a win-win agreement, the secretary was in an expansive, creative mode as he and his entourage made their way west to Yosemite. At Tuolumne Meadows they participated in the dedication of Mount Ansel Adams. Afterward, Hodel stretched his legs on a climb up Lembert Dome. On the descent, Hodel mentioned the Colorado dams and, according to David Brower, someone thrust David Cross's "Free the River" photograph of O'Shaughnessy Dam before him and asked if the secretary might wish to consider dismantling a larger dam. Little more was said, but in Sacramento Hodel called a meeting and Brower attended. The secretary's staff displayed the 1915 Hetch Hetchy Valley photographs by "Little Joe" LeConte. Hodel, remarkably, embraced the idea of restoration. It was Hodel's initiative, but of course Brower, who has often been described as a near clone of John Muir, loved the suggestion.[15]

As already noted, Hodel's proposal unleashed a flurry of activity. In Sacramento the lawmakers authorized $100,000 for the state Department of Water Resources to estimate the economic and environmental impacts of tearing down the O'Shaughnessy Dam. Some state senators, such as Quentin Kopp of San Francisco, believed Hodel's proposal was a "wacko idea."[16] Senator Milton Marks, also from San Francisco, saw the proposal as a thinly veiled first step to divert water from San Francisco to California Water Project customers in Southern California.[17] Generally the state legislators looked on the idea with a mixture of horror, curiosity, and amazement, for it was a thought that had never crossed their minds. However, in the Golden State any plan that changes water distribution attracts detractors and defenders. It seemed reasonable to at least study Hodel's idea, although many believed that California needed to be creating more water reservoirs, rather than tearing them down.

Interior Department officials took the secretary's idea seriously, although surely some privately questioned the wisdom of opening such a Pandora's box. The *Reno Gazette-Herald* commented: "It sounds crazy, doesn't it? Drain

Hetch Hetchy reservoir in Yosemite National Park to restore a valley for tourists—meantime wiping out the main water supply for San Francisco." Undeterred, Interior spokesman Tom Wilson affirmed that the department would "provide [Hodel] with what he wants as quickly as possible." He expected that within a couple of months the preliminary data could be in Hodel's hands, a report that would evaluate the possibilities of a much more detailed "feasibility study."[18]

While both the California state legislature and the Department of the Interior took action, the Sierra Club struggled with a position. On August 28, 1987, Sierra Club chairman Michael McCloskey, accompanied by members Sally and Les Reid, Carl Pope, and Joanne Hurley, drove to O'Shaughnessy Dam. Dean Coffey, superintendent of the Hetch Hetchy system and surely the most informed person regarding the consequences of removing the dam, met them. Coffey believed that some replacement water might be secured through building "wing dams" at the Don Pedro Reservoir. Furthermore, increasing the city's water rights at Don Pedro might be possible but would surely involve lengthy, costly legal fights. Coffey asserted that the existing system would not provide sufficient water if San Francisco elected to remove the dam. Furthermore, the Western River Guides, who had established a lucrative rafting business on the lower river, would object to removal. McCloskey noted that there was virtually no silt in the reservoir. This fact would accelerate restoration efforts should the dam be removed, but a siltless reservoir also offered a potent argument for San Francisco to maintain the dam, which Coffey believed could last a thousand years.[19]

McCloskey also met with Secretary Hodel in Washington, D.C. At the Sierra Club leader's request, the secretary gathered together Peter Koppleman of the Wilderness Society, Paul Pritchard of the National Parks and Conservation Association, Kent Olsen of American Rivers, Ed Ossan of the National Wildlife Federation, and Scootch Panhanian of the Western River Guides. It was a cordial meeting and the group focused on replacement water possibilities, including the Don Pedro Dam and the New Melones Dam on the Stanislaus River. They discussed San Francisco's water and power rights should the dam be breached. Although Hodel believed that the city had amortized its investment and benefited handsomely over the last 50 years, he still held that the city must be given some sort of settlement sum and, of course, replacement water. Probably most important, the group discussed political realities. Since Reagan would finish out his second term in January 1989, McCloskey urged the secretary to seize the opportunity while he was still in office to wrap up a study that could serve as the basis for a recommendation to Congress. He and the others agreed that this work should not be vested in any sort of commission, but should be done in-house. Hodel had some discretionary funds for preliminary work, although a full-fledged engineering feasibility study would require an appropriation from Congress.[20]

Evident at this meeting were the different agendas of the various environmental groups. The Western River Guides opposed raising the Don Pedro Dam since they would lose two miles of prime white water. Furthermore, breaching the O'Shaughnessy Dam would lead to an unregulated river, shortening the rafting season. The National Parks and Conservation Association worried that Hodel might use Hetch Hetchy restoration as a rationale for unwise development elsewhere. The National Wildlife Federation hoped to use this effort to force the secretary into improving the wildlife habitat in the Central Valley Project. In essence, all of these environmental groups supported the idea of restoration of Hetch Hetchy, but with their own conditions. All agreed, however, that the secretary should pursue a study immediately and that the National Park Service should play a key role. Furthermore, the study should not be publicized, and Congress should not be involved prematurely.[21]

Differences of opinion surfaced within the Sierra Club itself. Secretary Hodel counted on full support from the club, but it was not automatic. Harkening back to 80 years earlier, the Sierra Club Board of Directors hesitated to embrace a position that was unproven and unpopular in San Francisco. For David Brower, however, the dam should come down, and city politics, dollars, and budgets should not prevail over the rightness of an action. He was not above creating a clever message to enhance his position. On September 24, 1987, he wrote Sierra Club president Larry Downing and the Board of Directors, that he had received—on his computer screen—a letter from John Muir! Muir related to Brower that "the loss of the Hetch Hetchy Valley to the money-changers of San Francisco was one of the greatest disappointments of my life, and I saw little point in remaining on Earth after the Valley was lost." Now, with restoration possible, "it will be a glorious day when the recovery begins, and each succeeding day will be more glorious as each of the outcast [denizens] of the Valley, flower and trees and birds and other wildlife, return."

To board members suspicious of Hodel's motives, "Muir" had a message. "Would they rather curse a man for his demonic sins than reward him and all people who revere nature for his demonstrated virtue?" Muir (and Brower) clearly believed that the club should embrace the message and not be dissuaded by the messenger. In closing, Muir hoped that the present board would "renew the commitment to the ideal for which I founded the Club. It will hurt me greatly to be required to resign as Founder."[22]

The mystical Muir message forced the board to think in terms of history and commitment to an idea. However, it was ineffective. The board members refused to take a strong stand. The Hetch Hetchy Resolution, adopted on September 27, 1987, reaffirmed the club's opposition to the damming of the Tuolumne River. It also welcomed the initiative by Secretary Hodel and urged the secretary to "demonstrate the seriousness of his public proposal by assuring a feasibility study." In the endorsement, the board emphasized its "many and serious disagreements with Secretary Hodel and with the Administration," calling into question the sincerity of the proposal. It was a carefully worded, lukewarm endorsement of the secretary's initiative.[23] Brower was displeased. Here, finally, was the chance to right a wrong, and there should be no hesitancy. The club should act on principle and not be subject to monetary considerations, practicality, or the pulse of the public. This was not a matter for a Gallup poll to decide.

While the Sierra Club was unable to take a strong stand, no such problem plagued San Francisco. Removal of the dam would cost the city at least 50 percent of the output of the Moccasin and Kirkwood power houses which, in 1987, combined with the Holm unit to bring in approximately $25 million to the city's treasury. Such an amount of money paid for many services that the city could ill afford to lose. Mayor Feinstein and other city officials simply shook their heads in disbelief when confronted with the loss of income from the "vast hydroelectric power" system of Hetch Hetchy.[24] It was practically inconceivable that after so many years of trying to comply with section 6 of the Raker Act, the city had finally done so by the 1950s, only to have a significant portion of the power system dismantled by the interior secretary and his environmentalist supporters.

In the end, the mayor counted on the high cost of dam dismantling as the key to killing the idea. No reliable cost of the project existed, but Feinstein estimated $6 billion to $8 billion as not an unreasonable guess. All that money to resurrect 1.25 square miles of valley floor! She maintained that restoration "at the expense of the only water supply for two million people, 28 communities, thousands of businesses and two agricultural districts, and vast hydroelectric power—is not a good tradeoff." Mayor Feinstein implied, but never stated, that Congress would never authorize such a project. Legislators would be reluctant to vote such a sum to *build* a dam, let alone tear one down. Hodel, meanwhile, was well aware of the cost factor and had informed supporters that an $80 million appropriation might be possible, but not an $8 billion one.[25]

In November 1987 the Department of the Interior issued its report, entitled *Hetch Hetchy: A Survey of Water & Power Replacement Concepts.* It was a joint undertaking, with the Bureau of Reclamation examining the idea of restoring Hetch Hetchy Valley as a unit of the National Park Service. As the report subtitle indicated, the bureau focused on 11 potential replacement concepts should the dam be removed. The most practical matrices employed the idea of "reoperation" of the Tuolumne River system, acknowledging that the wild-and-scenic-river status of the lower Tuolumne would be impacted. Rerouting water through the city's tunnels and increasing storage at Don Pedro Reservoir could resolve the loss of water storage. The report noted that while the Hetch Hetchy reservoir could store 360,360 acre-feet, the Don Pedro's capacity was 2,030,000, nearly six times as much. The Bureau of Reclamation suggested that if the city made significant alterations, replacement water could be found:

> Based on dry-year hydrology, it appears that operation of the Hetch Hetchy system without Hetch Hetchy Reservoir could provide a minimum of 336,000 acre-feet of Tuolumne River water annually to San Francisco through the City's existing system. (At the present time, 336,000 acre-feet per year is the maximum amount that can be conveyed through existing conduits.) Over 100,000 acre-feet of replacement supply could be provided from river level diversion from below O'Shaughnessy damsite and another 250,000 acre-feet from changes in operation in Lakes Eleanor and Lloyd. In total, enough water could be captured to operate the City's existing conveyance system at full capacity and deliver almost 100,000 acre-feet more than is currently being used.[26]

While the discussion of water replacement was hopeful, the figures regarding electrical power were not. The city would lose capacity. Loss of the Hetch Hetchy Reservoir would leave only enough power to meet the city's uses. The Holm powerhouse (average annual generation 810 million kilowatt-hours [kWh]) would be unaffected, the Kirkwood powerhouse (615 million kWh annually) would lose much of its capacity, and the Moccasin Creek powerhouse (540 million kWh annually) would also lose some capacity. The surplus power sold to the Modesto and Turlock irrigation districts, of course, had become a "cash cow," significantly supplementing the city of San Francisco's budget. Removing the dam would result in a loss of millions of dollars in revenue, while the irrigation districts would also suffer a percentage loss of electricity deliveries from the Hetch Hetchy system. The Bureau of Reclamation projected that the cost of replacement power in the year 2000, at 8.47 cents per kWh, would be $109 million.[27]

Overall, the report was more upbeat regarding possibilities of replacement water and power than could have been expected. Perhaps it was the fact that the bureau's boss favored the idea. But more likely it reflected the slow evolution of the Bureau of Reclamation's mission in the West. Established in 1902 to build dams for irrigation and power purposes, that objective had been significantly diluted by the 1960s and 1970s. In the very year of Hodel's Hetch Hetchy idea, the bureau officially announced that its mission "must change from one based on federally supported construction to one based on effective and environmentally sensitive resource management."[28] The Bureau of Reclamation's assessment should be understood from this standpoint.

Written comments to the bureau's report were generally negative. The tally showed six agencies supported restoration, five had no position, and nine opposed it. Most of those supporting restoration were quite ambivalent and simply encouraged the consideration process to continue, specifically with a feasibility study rather than restoration itself. Those against tended to be much more detailed in their rejection of the study, and certainly more emotional. Letters addressed not only financial cost but also the projected loss of life involved in dismantling the huge dam. Perhaps the most determined opposition came from the San Francisco Bay Area Water Users Association, which considered itself a "co-grantee" with San Francisco under the Raker Act. The association membership included 30 cities and water districts in three Bay Area counties and consumed almost two-thirds of the Hetch Hetchy water delivered to the Bay Area. (In 1985–1986 the total delivery was 306,371 acre-feet, of which the users association consumed 195,457). In a 15-page single-spaced statement, President Warren Mitchell contested most of the bureau's findings, concluding, "We hope that, upon reflection, the Department will see no need to pursue the idea further." Mayor Dianne Feinstein agreed. "We cannot help but conclude," argued the mayor, "that the idea to remove Hetch Hetchy must rank at the bottom of the list of priorities for the federal government, the state and certainly the Bay Area." Those who opposed the restoration idea hoped that the report would put an end to what they considered a foolish scheme, perpetuated by radical environmentalists who had somehow hoodwinked the secretary of the interior.[29] With lukewarm backing at best, Hodel shelved the bureau's report, and like so many others, it faded from view.

The California state report, authorized by the state legislature, took an altogether different tack. *Restoring Hetch Hetchy* gave little emphasis to replacement of water and power, and more discussion and speculation on what might occur *if* the valley was restored to a natural state. The authors believed that with little human assistance, grasses would appear in a year or two. In regard to trees, "if Mother Nature does all the planting, then it would take much longer for a natural appearance." Trees would have to start as seeds, and not all the species originally in the valley would reestablish themselves. On the other hand, if humans assisted, moderately sized trees would be planted to replicate those growing in 1915, and within 10 years Hetch Hetchy would be quite presentable. Within 50 years the valley would appear natural to everyone except the experts. The report emphasized that restoration would be a "national experiment" that would draw attention across the country. Within 10 years the valley would be "fully usable" by recreationists.[30]

The cost of removal of the O'Shaughnessy Dam posed a problem. In the interest of economy the report suggested that workers might breach the dam, but not remove it. Jackhammers would reverberate throughout the valley as workers cut a tunnel through the base to allow drainage. Once accomplished, the tunnel could become the roadway into the valley. Planners, however, should not include intensive development as in Yosemite Valley, but rather place housing and tourist facilities downstream in the Poopenaut Valley. Shuttle buses might provide access, accommodating by 2000 one million visitors per year, or 2,700 per day on average, a far cry from the minuscule 110 visitors per day in 1987. Such a leap in visitation would equal about 15 percent of the 1987 visitors per day in Yosemite National Park.[31]

Eventually the state report had to abandon possibilities and address realities. The city's current diversion of 214,000 acre-feet of water from Hetch Hetchy Reservoir would drop to about 100,000, a serious loss. Regarding electricity, the report estimated that San Francisco would lose about $35 million per year if sold at 4.0 cents per kWh. These serious "takings" from the city could be ameliorated, but the six options suggested were complex, involving new construction of dams and tunnels, new irrigation agreements, and new power purchase contracts. None seemed satisfactory, although the most sensible was "to use the existing dams on the Tuolumne River and to construct a new dam that would flood the Poopenaut Valley, which is immediately downstream from Hetch Hetchy." Of course this smaller, less spectacular valley was within the boundaries of Yosemite National Park, too, but congressional legislation could transfer Poopenaut Valley to Stanislaus National Forest jurisdiction. The report concluded that both sides were right, but one side was just a little more right than the other:

> John Muir was right: the Hetch Hetchy Valley was gorgeous and would have provided incredible recreational opportunities. San Francisco was also right: the Hetch Hetchy Valley would make an exceptional municipal reservoir site. In 1913, Congress decided that San Francisco was "more" right and the Hetch Hetchy Dam was allowed. For the people

> of the 1980s, the issue of restoring Hetch Hetchy involves trade-offs between several very important public needs: recreation, aesthetics, high quality drinking water supplies, hydroelectric energy, and cost. We conclude that the existing Hetch Hetchy system is more valuable to society than a restored Hetch Hetchy.[32]

Earlier Secretary Hodel had written a supporter that the scenic wonders of Hetch Hetchy "need not be lost to the people forever. Maybe, with imagination, good will, and perseverance," mused Hodel, "we will be able to reclaim the national park land under the water of Hetch Hetchy Reservoir. Only [a] fair-minded study will tell whether this is an idea whose time yet may come." Now the interior secretary had not one, but two studies. Both were "fair-minded," objective efforts, and both indicated directly or indirectly that the time had not yet come for the great valley to be set free of the burden of almost 300 feet of water. Interior Secretary Hodel lit the fire in 1987; it blazed briefly, but then smoldered and died out. The Sierra Club moved onto other issues. The new George Bush presidency, with Manuel Luhan presiding as interior secretary, had no interest in pursuing Hodel's pet project.

Hodel had planted a seed, but maturation was not yet possible. However, the idea of renovation of landscapes has caught on, and local groups and national agencies have restored abandoned mining sites, toxic waste dumps, polluted wetlands, and other examples of our improvident use of the land. The idea of restoration of nature, of course, is not new. In his classic 1864 book, *Man and Nature,* George Perkins Marsh spoke of "restoration of disturbed harmonies." Marsh believed that worn-out lands could be restored. Human beings could become "co-worker[s] with nature in the reconstruction of the damaged fabric which the negligence or the wantonness of former lodgers has rendered untenantable."[33] One immediately thinks of American efforts to replant millions of acres of harvested timberland. Perhaps a more pertinent example was the attempt to restore topsoil in the Great Plains, so unwisely tilled by farmers in the 1920s and 1930s, the loosened soil lifted in great clouds by the winds of the West. No one would wish to indict the city of San Francisco with Marsh's idea of negligence or wantonness, but still the city had inundated a magnificent valley—many believed needlessly—and it would be intriguing to restore the "damaged fabric." It would be an experiment into the unknown, a challenge certainly not on the scale of the 1960s American decision to go to the moon, but path breaking in the sense that no humans have attempted the resurrection of nature on such a scale.

More recently, ecologist Frederick Turner has spoken to the general idea of restoration. He identifies the ecology movement in terms of (1) conservation, (2) preservation, (3) restoration, and (4) invention. In Turner's view restoration "seeks to reconstruct classical ecosystems, and is based on an assumption that to do so is not only possible, given nature's own easygoing and flexible standards, but also an important part of the human role within nature." To interact with nature and, if necessary, to assist in restoration defines our own humanness. Turner continues: "For the restorationist, nature is of intrinsic worth as well as valuable for human use."[34]

Ecologist Turner's fourth point, "inventionism," also has meaning for the submerged valley. Theoretically, he is convinced that ecologists cannot only conserve, preserve, and restore natural landscapes, but "when the occasion warrants and the knowledge is sufficient, . . . *create* new ecosystems, new landscapes, perhaps even new species." Human beings can be the "creator of natural diversity." The inventionist can "propagate life into presently dead regions of the universe, and even assist in the development of entirely new species."[35] Nature is constantly evolving, reinventing itself, and there is no reason for humankind to be excluded from the process. Rather clearly, Turner is thinking about the universe, but also earth.

Can the ideas of restoration work in Hetch Hetchy? Could the efforts of humans restore the "damaged fabric" of the valley? Could restoration aid humans just as much as nature? Would there be a role for the "inventionist" in the Hetch Hetchy Valley restoration? Many will argue no: Stay with the species of trees, shrubs, grasses, ferns, and riparian vegetation that existed in 1915. However, all of these questions would engender not only vigorous debate but also excitement, as botanists and plant biologists constructed a biologically harmonious ecosystem, working with nature but also shaping it. If we can accept the inventionist idea, the restoration of Hetch Hetchy could become a pathway to new knowledge, resulting in excitement between specialists, but within the general public as well. In an optimistic scenario, nature and ecologists would pool knowledge and creativity to form a sustainable relationship, creating a new and stunning landscape.

Lost in the historical debate, and usually in the contemporary one, has been the intrinsic value of nature, and particularly humans' ethical responsibility toward Hetch Hetchy Valley. John Muir understood it well, and yet he knew that Americans in his time were not prepared for what they considered foolish statements regarding the "rights of nature." If the valley was to be preserved, it would have to be based on human enjoyment. Today Americans are more comfortable with the idea of an expanded land ethic. But the question is: Are a sufficient number of Americans willing to grant Hetch

Hetchy new life based on an enlarged ethical responsibility? And if so, at what price? For those who believe in restoration, it does not matter. The monetary price tag is only one consideration. You cannot put a price tag—as Mayor Dianne Feinstein did—on all aspects of restoration. One earnest advocate, viewing the vast reservoir, observed that the valley had been "holding its breath."[36] Ken Brower, following in his father's footsteps, mused that "waiting in Yosemite National Park, under water, is a potential masterpiece of restoration, the recovery project to end all recovery projects; an enterprise which, if realized, would become a paradigm for all planetary restoration to come."[37] It was time to allow the valley to inhale deeply and come back to life. And, of course, for some devotees it is an ethical obligation to make this happen.

For the majority of Americans restoration of a valley and removal of a solid dam are ideas difficult to comprehend. Historically, we have built dams, not torn them down. Engineering journals have featured construction, not deconstruction. To do otherwise would be like a medical journal accentuating death rather than life. And yet many dams are reaching an advanced stage of life, for while they are neither sentient nor living, they do deteriorate. According to International Rivers Network, "by 2020 85% of all government owned US dams will be at least 50 years old, the typical design life span." Thus dams throughout the nation are—like the population—maturing. Some are losing their efficiency, costing too much in maintenance while holding back more silt than water. Many dams are in decline, and again, like medical doctors treating patients, hydraulic engineers spend their time checking on the condition of these structures to assure the public that their old age will not threaten downstream populations.

With the decline in the United States of dam building, the status of the civil engineer has lost some of its glamour. Most are maintaining rather than erecting new structures. They are more custodians than creators. We see few engineers with the visibility and heroic demeanor of George W. Goethals (Panama Canal), O'Shaughnessy, Mulholland, Holland, or Frank Crow (Hoover Dam). These men's structural works ushered in a modern world of abundance. Certainly they transformed the American West, but the status of their profession has changed. Just when that happened is difficult to say. For Mulholland, personally, it happened in 1928 when, on March 12 at 11:57 P.M., his St. Francis dam shuddered, cracked, and broke apart, sending a towering wall of water, mud, trees, and debris down a narrow canyon, eventually killing over 450 people.

The more general demise of dam building, however, came later than 1928.

By the 1960s most of the natural sites, such as Hetch Hetchy Valley, had been plugged. Government agencies began to look at cost-benefit analysis, asking questions about gains and losses in transforming a river and a valley. Of course such analysis was nothing new. The fight for Hetch Hetchy featured just about every conceivable pro and con of dam building. However, the public's *attitude* had changed as values shifted. Losses were magnified, and as John McPhee wrote in 1971, "there is something special about dams—as conservation problems go—that is disproportionately and metaphysically sinister." "Humiliating nature, a dam is evil—placid and solid."[38] Describing a dam in such anthropomorphic terms as *evil* and *sinister* seemed strange indeed, as did the concept of "humiliating nature." And yet, by 1971, with the environmental movement in full throttle, it seemed quite natural. After all, environmental groups had combined to defeat Bureau of Reclamation plans for a dam at Echo Park, within Dinosaur National Monument. And while the bureau was able to push through the Glen Canyon Dam, the last of the great dams in the West, in the mid-1960s it faced a bitter defeat in its attempts to build Marble Canyon and Bridge Canyon dams in the Grand Canyon. By 1987 the Bureau of Reclamation had revised its mission. No longer would it focus on construction of dams, but rather on water distribution efficiency and water quality. In essence, engineers would earn their living by manipulating the existing system in a more efficient manner.

While the pace of dam building has slowed to a feeble crawl, the idea of dam removal and land restoration has remained mainly a concept—not a reality. However, at least two examples are instructive. The breeching of the Edwards Dam on July 1, 1999, was a significant event. For the first time since 1837 the waters of the Kennebec River in Maine ran free. The decommissioning and destruction of the two-story, 900-foot-long dam represented the culmination of a long process. In 1986 the Federal Energy Regulatory Commission (FERC) refused to renew the Edward Manufacturing Company's 50-year license to operate the dam, ordering the company to shut down its turbines, tear down the dam, and return the river to its free-flowing state. Thirteen years of negotiation followed. When the breeching finally occurred, it took on the character of a celebration and a party. The curious early birds arrived at 6:00 A.M., the first of more than a thousand spectators. Helicopters circled and newspaper photographers aimed their long-lensed cameras at a 60-foot gap in the dam, where once workers removed the cofferdam barrier, the river would flow free. Speeches were made and a generally festive atmosphere prevailed. One group seemed more reserved. In a single row sat Rebecca Wodder and Margaret Bowman, representing American Rivers; Amos Eno of the National Fish and Wildlife Foundation; Todd Ambs of the River Alliance of Wisconsin; Mike Lopushinsky of New York Rivers United; and Pam Hyde from the Glen Canyon Institute. "They sat on folding chairs," observed John McPhee, "as if on a ship's deck for a formal surrender. For the victors, this was Yorktown, Cornwallis sulking in his tent."[39] At the proper moment, they felt satisfaction as the bulldozers removed the cofferdam and the river ran free. Obviously, the triumph was a joint effort of many persons and many environmental groups. It was the first major dam in the nation to be ordered out of existence. Today, both the salmon and the natural environment have returned to the Kennebec River. It does not take long for nature to reclaim its lost habitat.

In the Pacific Northwest a similar story is taking place on the Elwha, a major river on the Olympic Peninsula. During the same era when San Francisco acquired the Hetch Hetchy Valley, the Olympic Power Company, owned by entrepreneur Thomas Aldwell, constructed the Lower Elwha Dam, bringing much-needed power to the small town of Port Angeles. In 1916 Aldwell convinced the Zellerback family of San Francisco to purchase his company and also to build a paper plant at Port Angeles, ensuring jobs and a future for the struggling town. In the late 1920s the Crown Zellerback Corporation built the Glines Canyon Dam, a much larger dam further upriver. The two dams absolutely killed the vast salmon and steelhead runs on which many people depended, particularly the members of the Klallam Indian tribe. Yet the political power of Crown Zellerback and the jobs the plant created overwhelmed any protests. In the mid-1970s Crown Zellerback, like the Edwards Company, had to appear before the Federal Energy Regulatory Commission for renewal of its 50-year license. In an earlier era such an application would have been approved automatically. But times were changing. Complicating the situation was the fact that, like Hetch Hetchy, both the Elwha and the Glines Canyon dams were within a national park. However, in this case the dams existed before Congress established Olympic National Park in 1938.

Again, as with Edwards Dam, the commission had to weigh benefits and detriments. In FERC's deliberations the benefits seemed rather minuscule. The 20 megawatts of power that the two dams churned out appeared paltry, and the loss of that power could be made up primarily through conservation—in effect, more efficient consumption. There were few other arguments for the dams. More and more locals, and certainly the Klallam tribe, considered the return of salmon runs more important than electric power. FERC refused to extend the license. Again, years of litigation and negotiation ensued, but in 1992 the Elwha River Restoration Act passed Congress on the last day of the legislative session. A reluctant President

George Bush signed the bill into law. The act authorized the Department of the Interior to purchase the two dams for not more than $29.5 million, with the intent of removing them and returning the river to its normal flow.[40] It is one thing for Congress to authorize an expenditure, and another to fund it. In February of 2000 the government finally found the money to purchase the two dams. As of this writing the two Elwha River dams remain intact, but it is only a matter of time before millions of salmon once again churn up the unobstructed river.

In both the Elwha River and Kennebec River cases one cannot ignore the crucial role of the Federal Energy Regulatory Commission. When Congress established it under the Federal Power Act of 1920, the commission was beholden to the power interests. As one might expect, it established hydropower as the "highest and best use of watersheds." But in both the Kennebec and Elwha River cases, an amended Federal Power Act instructed FERC to reestablish its priorities. The commission was to give "equal consideration" to wildlife, recreation, environmental quality, and related factors.[41] Weighing the values of power production and corporate profit with the Edwards Dam, the commission tipped over to the environmental side. A similar situation occurred with the Elwha River dams when the draft environmental impact statement, required by FERC, made salmon restoration its top priority. To renew its license, Crown Zellerback would have to restore the salmon to the river—something the company could not accomplish.[42]

This discussion is not meant to suggest that the same principles apply to Hetch Hetchy. The Federal Energy Regulatory Commission has no jurisdiction over the O'Shaughnessy Dam, authorized seven years before the 1920 law was enacted. The key point is that major dams have, and will, be breached and the preponderance of public opinion was that their time had come. Like an old automobile, these dams were worn out and no longer fulfilled their purpose. Both the Matilija Dam in Southern California and the Elk Creek Dam (never completed) no longer serve any useful function.[43] But like old cars, most of the thousands of "elderly" dams across the country will be repaired and reinforced, rather than removed. Yet the FERC actions mean that that no longer can dam owners continue with business as usual. Interior Secretary Bruce Babbitt, speaking at the Kennebec River in 1999, sounded a cautionary note: "This is not a call to remove all, most, or even many dams. But this is a challenge to dam owners and operators to defend themselves, to demonstrate by hard facts, not by sentiment or myth, that the continued operation of a dam is in the public interest, economically and environmentally."[44]

Throughout the 1990s, aside from the lone voice of David Brower, water and power flowed to San Francisco without criticism or distraction. As energy prices increased and water became increasingly valuable, the Hetch Hetchy system became the city's greatest asset, funding police protection, city park maintenance, and various social services. Hodel's wicked idea became only a distant memory. Yet in the San Francisco Sierra Club headquarters the fire had not totally gone out. Carl Pope, who had been a leader in the club's effort to support Hodel, asked a young staffer, Ron Good, to study a pile of documents on Hetch Hetchy. Nothing came of his work, but a seed had been planted. Good left the San Francisco office in 1989 for Ohio, but then returned to Southern California to teach school. He could not escape Hetch Hetchy, and he soon composed a 20-page treatise expressing his thoughts on the lost valley. He proposed the formation of a restore–Hetch Hetchy task force to be funded by the Sierra Club Foundation, the club's tax-deductible arm. He took a job in Yosemite Valley and in his spare time volunteered at the Sierra Club's handsome LeConte Lodge. One day Harry Wilkinson, a retired professor from Rice University, visited the lodge and in the course of his conversation with Good, Hetch Hetchy Valley became a subject. Soon Wilkinson wrote out a $1,000 check to the Sierra Club Foundation.

The money was a catalyst, for the task force members found that the Sierra Club directors and staff did not view Hetch Hetchy as a high priority. They were not hostile, but neither would they move the club toward a greater commitment. Given the situation, the task force—well aware that John Muir and those willing to fight for Hetch Hetchy Valley in 1909 had split off from the club to form the Society for the Preservation of National Parks—moved to end its direct affiliation with the Sierra Club. By October 2000 the directors of the newly named "Restore Hetch Hetchy" (RHH) task force filled out the paperwork and received the federal 501C3 nonprofit status. A tiny organization needs a benefactor, and RHH found it in the California Planning and Conservation League, particularly in Gerald Meral, who joined the RHH board. Meral was a veteran of California water and dam fights, including that over the New Melones Dam on the Stanislaus River. His support and those of other board members, such as Brower, meant the RHH was off to a good start. The following July, the board appointed Ron Good as executive director.[45]

The specter of the tiny Restore Hetch Hetchy group (with about 1,500 members) taking on San Francisco and its Public Utilities Commission is more than a little reminiscent of Muir, Colby, McFarland, and Robert Underwood Johnson and their very unequal fight. Working with only a small budget and relying primarily on volunteer help, Restore Hetch Hetchy must fight a difficult battle. Like a tiny gnat, it can only irritate the giant. Yet like so many small, underfunded environmental groups of the past century, RHH depends on what it views as the rightness of its cause. In his last public appearance a wheelchair-bound David Brower spoke to the Yosemite Association Annual Meeting in September 2000 of his commitment to the restoration of the Hetch Hetchy Valley. With such recognized figures as Brower, RHH has won both publicity and converts.

The group aims to win grassroots support as Good and others speak to environmental and community groups across the country, asking that they pass resolutions supporting the concept of restoration. They take the view that a win-win solution is possible, with the Yosemite National Park winning back one of its most scenic areas while the city will avoid forfeiting its water supply. Admittedly, under most scenarios the city will lose significant electrical generating capacity, a serious issue in a state that recently experienced an eye-opening electric energy crisis. But RHH is quick to point out that the loss will be well under 1 percent of California's production and can be recouped through wind and solar facilities and by energy conservation. In regard to cost RHH maintains that one cannot put a price on beauty and that some of the restoration costs must be borne by all of the American people. Aesthetic and ethical considerations must take precedence over those of money and engineering.

What are the chances of restoration of the Hetch Hetchy Valley—America's Lost Valley? Candidly, they are slim. San Francisco takes little notice, and with the passage of the $1.6 billion Hetch Hetchy Capital Improvement bonds in the fall of 2002, the San Francisco Public Utilities Commission is prepared to move ahead with a number of projects to upgrade the system. The 30 participating wholesale customers (Bay Area cities in Alameda, San Mateo and Santa Clara counties) will contribute $2 billion to this effort, thus creating an enormous purse of $3.6 billion. However, O'Shaughnessy Dam will receive very little attention, given that it is in excellent shape for a 70-year-old dam. Most of the money will be directed toward 77 projects involving aging aqueducts, tunnels, and replacement of old Spring Valley components still in use. The commission plans to spend much of its public funds to alleviate earthquake vulnerabilities as well as to increase the size of Calaveras Reservoir from 30,000 acre-feet of storage to over 400,000. All of this will cost the Hetch Hetchy water user plenty. Today the average water bill in San Francisco is $14 a month, but by 2015 it will increase to $41. A typical restaurant owner would see his or her bill increased from $314 to $889.[46]

Proposition A, as the city designated the Hetch Hetchy bond issue, was not universally popular. Businesses were alarmed by the increases, thinking that the hotels, landlords, and restaurants would be paying far too much. The San Francisco Bay Chapter of the Sierra Club opposed the proposition. The main concern was that San Franciscans would be paying for sprawling development in the outlying towns. Other political issues arose with Proposition A, but none involved the question of the O'Shaughnessy Dam or the submerged Hetch Hetchy Valley. In the final tally, voters approved the bond issue, 53 percent to 47 percent, not an overwhelming vote of confidence.[47] Sixty or seventy years ago, when passage of Hetch Hetchy bonds required a two-thirds vote, the Municipal Utilities District would have been seeking other funding options.

Between passage of Proposition A and the Public Utility Commission's preoccupation with other matters, it is evident that San Francisco has little interest in redesigning a system in order to allow for the restoration of Hetch Hetchy Valley. The commission has no subcommittee on Hetch Hetchy's restoration, and it engaged in no discussion about the proposal during the debate on Proposition A. Hetch Hetchy restoration is not a major issue around City Hall, although three supervisors did introduce a resolution for a feasibility study. It went nowhere. On the federal level, if restoration receives an audience, the city can be assured that now senator Dianne Feinstein will fiercely oppose and kill any federal support for such a plan. When she was mayor, Feinstein admitted that the building of O'Shaughnessy Dam may not have been a good idea. However, that admission has not translated into support for removal. Feinstein believes that water quality would diminish, electricity production would be seriously lessened, and the cost of replacement and demolishing the dam would be too high, "with estimates for the project close to $3 billion." Such resources could be better used, she argues, "to help California plan for its future water needs."[48]

The National Park Service is quite ambivalent. Privately many Yosemite Park officials express pleasure at the thought of restoration. Furthermore, dedicated former superintendents such as Michael Finley and David Mihalic have expressed support for the idea. Yet the Park Service has no desire to pick a fight with the city. In the spring 2001 *Yosemite Guide*, the handout given to visitors at the entrance stations, touted Hetch Hetchy as a place of "beauty and solitude"—the "quiet corner" of the park, which is well worth the drive.[49] Furthermore, in an ironic twist, both the city and the park take pride in noting the designated wilderness areas that border the reservoir. Many park personnel would probably relish restoration both as historical retribution and as an ecological experiment. But as yet, aside from Hodel, persons in power have not signaled their support.

FIGURE 27. When the city occasionally draws down the water, the reservoir is not a pretty sight. The photographer, Carmen Magana, took this shot in 1993. *Courtesy of the SFPUC.*

FIGURE 28. When the reservoir is full and water is flowing over the drum gates, the scene approaches what O'Shaughnessy hoped: the blending of technology and nature. *Courtesy of Dan Flores.*

It will take time and much education for the American public to grasp Ken Brower's idea that Hetch Hetchy provides "a potential masterpiece of restoration," especially in the absence of a charismatic environmental leader, a modern-day John Muir. Moreover, there are powerful arguments for retaining the O'Shaughnessy Dam and the reservoir. These economic and engineering arguments are compelling, and could be successfully challenged only through a long-term process of collecting a constituency around acceptance of the position that nature has an intrinsic right to survive and even a right to rebirth.

As Aldo Leopold's concept of human ethical responsibility to the land takes hold with more and more Americans, restoration may eventually seem more a duty than a preposterous idea. But that day is not yet here. Such sentiments, if they do develop, are likely to arise in places other than San Francisco. In Muir's day he and others were able to rally opposition to flooding Hetch Hetchy Valley from one end of the nation to the other. San Franciscans became particularly enraged when Easterners presumed to tell the city what to do. Today, it is not altogether different. In October 2002 the influential *New York Times* ran a page 1 story on the restoration idea. It featured Donald Hodel, living in semiretirement but still committed to his idea advanced 14 years earlier. He likened the argument that San Francisco had a "birthright" to flood the valley with the "arguments made by slaveholders in opposition to abolition." The analogy was surely shaky, but nevertheless memorable. Four days later the *Times* ran an editorial, "Bring Back Hetch Hetchy?" concluding that "the least we can do is endorse a feasibility study. It may well lead to something remarkable." The *San Jose Mercury News* agreed in an editorial titled "Sure, Study It."[50] Former mayor Willie Brown and the city supervisors ignored this advice.

Mayor Brown has had some of his own ideas regarding Hetch Hetchy. In 1998, when he dined at a San Francisco bistro, he could not help but notice high-priced bottled water from foreign lands. Why not, he thought, bottle tasty Hetch Hetchy water? "Hetch Hetchy has a cachet," announced the mayor, adding that "everyone will pay for this water. It is worth drinking." By 2003 the city began to market the water in a limited way. It seemed half joke, half publicity stunt, but raised the ire of many. Ron Good took the opportunity to state that this was just one more example "of San Francisco exploiting the natural resources of Yosemite National Park." A park ranger commented that the water could be a reminder that perhaps Hetch Hetchy should never have been dammed. However, the most stinging cut of all came from the *Los Angeles Times,* which loved to tweak its northern neighbor: "Shame. Hetch Hetchy may be as fresh and tasty as bottled water gets, but any good environmentalist with a sense of history would rather drink irrigation runoff." The *Times* editorial confessed to Los Angeles's own rather unethical drive for water, but in recent years "Los Angeles has given up—grudgingly, of course—much of its Owens water to restore Mono Lake and put water back in the Owens River Gorge. Someday," the editorial suggested, "perhaps San Francisco will recognize that its pride in Hetch Hetchy is misplaced and that dismantling the dam is something that is really worth San Francisco's image of itself."[51]

The drama will, of course, go on. While San Francisco may choose to ignore the idea, dam restoration has become an option.[52] Even in Yosemite Valley the National Park Service in November 2003 removed the Cascades

Dam, a small structure built in 1917 on the Merced River to generate electricity.[53] Elsewhere dam demolition is a widely discussed topic. Perhaps symbolic in its title, the Aspen Institute recently sponsored a conference entitled, "Dam Removal—A New Option for a New Century."[54]

Although in denial, San Francisco, a self-proclaimed wellspring of environmental passion, will continue to wrestle with its conscience. Perhaps in an effort to awaken that conscience, in February 2004 President George W. Bush's budget proposed to raise the rent the city has been paying for over 70 years from the current $30,000 per year to $8 million per year. Senator Feinstein announced her opposition, but Congressman George Radanovich, a Republican from Mariposa, California, and chair of the National Parks Subcommittee, admitted that there should be an end to San Francisco's "overly advantageous deal."[55]

Whatever the outcome of this latest dispute, the restoration of Hetch Hetchy Valley holds a certain attraction. The Bureau of Reclamation's 1987 report caught that spirit of excitement:

> Initially startling, this idea [of restoration], on second consideration, begins to intrigue the mind and free the imagination to consider the creative potential of such a proposal. One begins to see the possibilities inherent in the opportunity to re-evaluate a past decision in the light of not only today's but tomorrow's needs.[56]

As we move into the new century, environmental groups will continue to agitate for change, reminding those in power of the mistakes of history. Defenders of the O'Shaughnessy Dam will decry suggestions of removal as acts of sacrilege, while others will declare such action as an act of salvation. As in the past, Hetch Hetchy will continue as a touchstone, reflecting the needs of San Francisco and society but also acting as a barometer of environmental attitudes.

Afterword

IN JULY 2003 three friends and I hiked from Tuolumne Meadows to the Hetch Hetchy Reservoir, ending our seven-day odyssey at the O'Shaughnessy Dam. The trail works its way down the Grand Canyon of the Tuolumne River to Pate Valley, then swings up Piute Creek to Pleasant Valley, tops out at Rancheria Mountain, and then descends in a long roller-coaster walk to Rancheria Falls and the reservoir.

Of course, we talked rather endlessly about Hetch Hetchy, as we descended some 4,000 feet along the Tuolumne River, a stream that rarely pauses from its wild rush down the canyon. For 20 miles we exulted in white water rushing over polished granite on its way to the sea. One of our party brought along a copy of Wallace Stegner's brief essay *The Sound of Mountain Water* and read a portion as we paused on a glistening granite slab to view LeConte Falls:

> By such a river it is impossible to believe that one will ever be tired or old. Every sense applauds it. Taste it, feel its chill on the teeth: it is purity absolute. Watch its racing current, its steady renewal of force: it is transient and eternal.[1]

Of course, the Tuolumne River water, like most California streams, is consumed by our insatiable thirst well before it reaches the sea, its natural destination. But the Grand Canyon of the Tuolumne is a magnificent slice of nature. It features not only rushing water and occasional deep green pools but polished granite everywhere, the result of glacial action thousands of years ago. The slopes, cliffs, and domes reflect light, and if John Muir had not called it "The Range of Light" over a hundred years ago, we would have come up with the description ourselves. Many years ago Muir, who surely was a travel guide as well as a naturalist-philosopher, urged everyone to take this hike. He believed that "every one who is anything of a mountaineer should go on through the entire length of the cañon, coming out by Hetch Hetchy. There is not a dull step all the way."[2] We could not have agreed more. William Badè, Muir's young friend and admirer, struck off with friends down the canyon 100 years ago, admiring "a river capable of such acrobatics feats, such impetuous *abandon,*" one that "responds to the pull of gravity with almost incredible momentum." Of course they did not have the benefit of a trail. As he tells it, "one of our number lost his footing on a narrow ledge four hundred feet above the churning river, and would have lost his life had he not caught with his left hand a tough-rooted young oak that grew in a crevice." We did not experience such hair-raising scrapes, but neither did we have the pleasure of entering the Hetch Hetchy Valley from what Badè called "the natural entrance or exit of the Tuolumne Canyon." We were further denied Badè's experience of passing through "the portals of Hetch Hetchy, next to Yosemite the greatest natural cathedral on the Pacific Coast." One imagines silence, but Badè, a scholar and theologian, wrote that "from the richly carved choir galleries came the joyous music of many waters, and the deep organ tones of full-throated waterfalls pealed forth ever and anon as we threaded its aisles on subsequent days."[3] Badè, like so many other visitors, lingered in the deep granite basin echoing with flowing waters.

What we experienced was quite different. After camping at Rancheria Creek, we paused high above the reservoir on our way to Wapama Falls. The waters reflected the sun, quiet and peaceful. No motorboats or water skiers intruded on our thoughts. The reservoir was full, so our view included no bathtub rings. Perhaps Phelan, Freeman, and O'Shaughnessy were right. Perhaps the reservoir represented the perfect blending of technology and nature. On a windless, cobalt blue–sky day, looking down on a serene mountain lake enclosed by granite cliffs, one could not be too angry over what had happened 80 years earlier.

As I viewed the scene, I thought about the meaning of the Hetch Hetchy fight. Was this the victory of public power over private utility interests? Did the reservoir represent the triumph of utilitarian scientists and engineers over the "nature lovers"? Was it all about the imperial city of San Francisco's dominance over the rural interests of the Central Valley? Did the valley represent the touchstone of variations of Progressive principles? Would the San Francisco Bay Area have attained its current prosperity and growth without the shining reservoir below? Was this an issue of California as wilderness, or California as suburb and city? Was this an issue of class, with the more refined, educated, white Americans aligned against the teeming masses of variegated immigrants who needed water in crowded San Francisco? Would the San Francisco Bay Area been denied a pure supply of water if the river still flowed through the canyon below?

All of these questions I answered with a qualified no—qualified because complicated questions require nuanced answers. My study revealed complexity, and the reasons for favoring or opposing the Hetch Hetchy system were as varied, and as shifting, as the currents of the river. In the end, however, I concluded that if forced to identify one overreaching reason for the dam's existence, it was a failure of the democratic process or, perhaps more accurately, the bias of the democratic process toward San Francisco's power and wealth. Serene as the reservoir appeared at the base of Kolana Rock, it should not be there. I should not have to imagine the valley that Muir first described. It should still exist. Why did it not? Arguments that San Francisco and its surrounding communities would languish without the reservoir were spurious. No one should have believed that the future growth of the San Francisco Bay Area depended on the building of the Hetch Hetchy system. At least three other Sierra Nevada sources of quality water were available, all at nearly the same cost and free of the encumbrances of a national park. But the politicians of San Francisco, with wealth, power, and sympathy engendered by the 1906 earthquake and fire, insisted on the invasion of the Hetch Hetchy Valley. In the sense of the tyranny of the rich and powerful, the democratic process failed. The valley was lost to money, lobbying, and political pressure that only one side could muster. By legislative caprice and power politics the city got its valley.

But my somber mood lightened as I looked on the reservoir with greater optimism. In spite of protestations to the contrary, here was a fight in which there were no real villains. Perhaps there were highly emotional nature lovers and arrogant engineers, but both sides were well-meaning and surely embraced the Progressive ideas of their day.[4] Rather than wish the reservoir away, maybe one should celebrate that the valley was *nearly saved*. After all, many other impressive valleys had been and would be dammed without a word of protest. Here at Hetch Hetchy, a new and different constituency had made a stand, and although the odds against it were overwhelming, it came close to winning. Perhaps the reservoir represented a martyrdom to an ideological cause that gained momentum in the new century.

Walking on to Wapama Falls and a welcoming swim in the noonday heat, I speculated whether the O'Shaughnessy Dam would last a thousand years. Perhaps it would, but it will not last forever. Human accomplishments are so transient, so ephemeral, so insignificant. Through at least 99 percent of its existence, the valley teemed with life. I felt sure it would again.

NOTES

INTRODUCTION

1. Ray W. Taylor, *Hetch Hetchy: The Story of San Francisco's Struggle to Provide a Water Supply for Her Future Needs* (San Francisco: Ricardo J. Orozco, Publisher, 1926), vii.

2. Articles include Kendrick A. Clements, "Politics and the Park: San Francisco's Fight for Hetch Hetchy, 1908–1913," *Pacific Historical Review* 48 (May 1979), 185–215; Elmo R. Richardson, "The Struggle for the Valley: California's Hetch Hetchy Controversy, 1905–1913," *California Historical Society Quarterly* 38 (1959), 249–58. Significant chapters or entries in books include Michael L. Smith, *Pacific Visions: California Scientists and the Environment, 1850–1915* (New Haven: Yale University Press, 1987), ch. 8; Roderick Nash, *Wilderness and the American Mind* (New Haven: Yale University Press, 1967), ch. 10; John Ise, *Our National Park Policy: A Critical History* (Baltimore: Johns Hopkins University Press, 1961), 85–96; Alfred Runte, *National Parks: The American Experience* (Lincoln: University of Nebraska Press, 1979), 77–81; Norris Hundley, *The Great Thirst: Californians and Water, 1770s–1990s* (Berkeley and Los Angeles: University of California Press, 1992), 169–200; Elmo Richardson, *The Politics of Conservation: Crusades and Controversies, 1897–1913* (Berkeley and Los Angeles: University of California Press, 1962), 72–84. One must, of course, include Holway R. Jones's *John Muir and the Sierra Club: The Battle for Yosemite* (San Francisco: Sierra Club, 1965).

3. Linnie Marsh Wolfe, *Son of the Wilderness: The Life of John Muir* (New York: Alfred A. Knopf, 1945, repr. Madison: University of Wisconsin Press, 1978), 154.

1. THE USES OF THE VALLEY

1. John Muir, "Hetch Hetchy Valley," *The Overland Monthly*," 11 (July 1873), 46. Muir and many other placed a hyphen between *Hetch* and *Hetchy*. In the twentieth century most authors omitted the hyphen. For consistency and to eliminate confusion, I have chosen to use only the unhyphenated version, "Hetch Hetchy," throughout this book.

2. Josiah Dwight Whitney, *Yosemite Guide-Book: A description of the Yosemite Valley and the Adjacent Regions of the Sierra Nevada, and the Big Trees of California* (published by authority of the California State Legislature, 1869) 83–84.

3. John Muir, "The Hetch Hetchy Valley," *Boston Weekly Transcript*, March 25, 1873.

No doubt Muir took some pleasure in publishing his Hetch Hetchy findings in a paper read by Whitney and his scientific friends.

4. Muir, "Hetch Hetchy Valley," *Overland Monthly,* 42. Interestingly, Whitney accepted the glacial origins of the Hetch Hetchy Valley while denying significant glacial action in Yosemite. In *The Yosemite Guide-Book* (111) he states, "There is no doubt that the great glacier, which, as already mentioned, originated near Mt. Dana and Mt. Lyell, found its way down the Tuolumne Cañon, and passed through the Hetch Hetchy Valley."

5. James P. Walsh, Timothy J. O'Keefe, *Legacy of a Native Son: James Duval Phelan and Villa Montalvo* (Saratoga, Calif.: Forbes Mill Press, 1993), 81–87.

6. Samuel Taylor Coleridge, "Rime of the Ancient Mariner" (1797), part 2.

7. Quoted in Emily Wortis Leider, *California's Daughter: Gertrude Atherton and Her Times* (Stanford, Calif.: Stanford University Press, 1991), 257.

8. Because there is very little research specifically about the Indians who inhabited the Hetch Hetchy Valley, I have relied on what is known of the Central and Southern Miwok and the Paiute of the Mono Lake area, who largely inhabited and/or visited Yosemite Valley. The two valleys were, of course, very similar in origin, vegetation, and altitude and only twenty miles apart. See Craig D. Bates, Martha J. Lee, *Tradition and Innovation: A Basket History of the Indians of the Yosemite–Mono Lake Area* (Yosemite National Park: Yosemite Association, 1990).

9. See Henry T. Lewis, "Patterns of Indian Burning in California: Ecology and Ethnohistory," in *Before the Wilderness: Environmental Management by Native Californians,* ed. Thomas C. Blackburn and Kat Anderson (Menlo Park, Calif.: Balleria Press, 1993), 55–79.

10. Helen McCarthy, "Managing Oaks and the Acorn Crop," in *Before the Wilderness,* 222.

11. Ibid., 214–17.

12. Described in ibid., 218–20.

13. Peter Browning, *Yosemite Place Names* (Lafayette, Calif.: Great West Books, 1988), 60; "Yosemite Guide," 30 (Spring 2001), 1.

14. Irene D. Paden, Margaret E. Schlichtmann, *The Big Oak Flat Road: An Account of Freighting from Stockton to Yosemite Valley* (San Francisco, 1955), 191.

15. Browning, *Yosemite Place Names, 60–61;* Ray W. Taylor, *Hetch Hetchy: The Story of San Francisco's Struggle to Provide a Water Supply for Her Future Needs* (San Francisco: Ricardo J. Orozco, Publisher, 1926), 37. This was the first published book on the Hetch Hetchy issue. Reflecting the views of San Francisco politicians and engineers, it lacks critical perspective.

16. Margaret Sanborn, *Yosemite: Its Discovery, Its Wonders, and Its People* (New York: Random House, 1981), 13.

17. Jean-Nicolas Perlot, *Gold Seeker: Adventures of a Belgian Argonaut during the Gold Rush Years* (New Haven: Yale University Press, 1985), 282–84.

18. Paden, Schlichtmann, *The Big Oak Flat Road,* 188–90.

19. Taylor, *Hetch Hetchy,* 37.

20. Muir, "The Hetch Hetchy Valley," *Boston Weekly Transcript.*

21. As Muir so eloquently described it in *My First Summer in the Sierra,* he was

directly involved in tending sheep in 1869. However, there is no evidence that either he or the sheep entered the Hetch Hetchy Valley. One can suppose, however, that he first heard of Hetch Hetchy during that memorable summer.

22. Whitney, *Yosemite Guide-Book,* 111–12.

23. Muir, "Hetch Hetchy Valley," *The Overland Monthly,* 45.

24. J. M. Hutchings, *In the Heart of the Sierras: The Yosemite Valley* (Oakland: Pacific Press, 1886). Hutchings includes a chapter on the Big Oak Flat Road, but never mentions the Hetch Hetchy Valley.

25. Mrs. Carr to John Muir, fragment of letter, circa October, 1872, in John Muir Papers, reel 2, Correspondence, 1969–1873.

26. Albert Bierstadt to Mrs. Sawyer and Mrs. Williston, November 8, 1976, in Mount Holyoke College Library/Archives.

27. Probably the most impressive (37 in. ? 58 in.) Bierstadt painting of the Hetch Hetchy Valley hangs in the Wadsworth Atheneum gallery in Hartford, Connecticut.

28. Information taken from a memo from Wendy Watson, Mount Holyoke College Art Museum, to Ron Good, August 31, 1998.

29. Notes from Alfred Harrison, Jr., The North Point Gallery, in the Restore Hetch Hetchy Archive.

30. Keith published a riveting account of this adventure with Muir in the Boston *Advertiser,* 1874. Keith's biographer Brother Cornelius reproduced the piece in *Keith: Old Master of California* (New York: G. P. Putnam, 1942), 61–65.

31. Ibid., 477.

32. Alfred Runte, *Yosemite: The Embattled Wilderness* (Lincoln: University of Nebraska Press, 1990), 9–12. Runte suggests that the name *Yosemite,* rather than a word for grizzly bear, is a corruption of *Yo-che-ma-te,* an Ahwahneechee word for "some among them are killers," no doubt referring to the activities of a good number of the white militia members. See Bates, Lee, *Tradition and Innovation,* for information on the evolution of the Yosemite Valley Indians in the twentieth century.

33. Hans Huth, *Nature and the American: Three Centuries of Changing Attitudes* (Lincoln: University of Nebraska Press, [1957] 1972), 148–49.

34. Muir noted on his 1871 visit to Hetch Hetchy the presence of two shacks, which he assumed were summer shelter for shepherds.

35. Robert Underwood Johnson to John Muir, December 13, 1877, in John Muir Papers, reel 3, Correspondence, 1874–1879.

36. Robert Underwood Johnson, *Remembered Yesterdays* (Boston: Little, Brown and Co., 1923), 284.

37. The two Muir articles were "The Treasure of the Yosemite," *Century Magazine* 60 (August 1890), 483–500, and "Features of the Proposed Yosemite National Park," *Century Magazine* 60 (September 1890). 656–67.

38. Runte, *Yosemite,* 53–55. Also see Richard Orsi, "'Wilderness Saint' and 'Robber Baron': The Anomalous Partnership of John Muir and the Southern Pacific Company for Preservation of Yosemite National Park," *Pacific Historian* 29 (Summer–Fall 1985), 136–56.

39. The mixture of private and public lands was certainly not ideal from a man-

agement or conservation point of view. Yet it is a reality that almost all national parks and national forests must face. For San Francisco it proved to be advantageous.

40. Roderick Nash, "Conservation as Anxiety," in *American Environmentalism: Readings in Conservation History,* ed. Roderick Nash (New York: McGraw-Hill, 3rd ed., 1990), 105–12.

41. Peter J. Schmitt, *Back to Nature: The Arcadian Myth in Urban America* (Baltimore: Johns Hopkins University Press, 1969, repr. 1990), 156.

42. Robert Marshall, "The Problem of the Wilderness," *Scientific Monthly* 30 (February 1930), reprinted in J. Baird Callicott, Michael P. Nelson, eds., *The Great New Wilderness Debate* (Athens: University of Georgia Press, 1998), 85–102.

43. See Laura and Guy Waterman, *Forest and Crag: A History of Hiking, Trail Blazing, and Adventure in the Northeast Mountains* (Boston: Appalachian Mountain Club, 1989), 307–8. The best books on this "back to nature" movement include Nash, *Wilderness and the American Mind;* Schmitt, *Back to Nature;* David E. Shi, *The Simple Life: Plain Living and High Thinking in America* (New York: Oxford University Press, 1985); and Stephen Fox, *John Muir and His Legacy* (New York: Little, Brown, 1981). Also see Ralph H. Lutts, *The Nature Fakers: Wildlife, Science and Sentiment* (Charlottesville: University Press of Virginia, 2001).

44. Quoted in James Mitchell Clarke, *The Life and Adventures of John Muir* (San Francisco: Sierra Club Books, 1980), 249.

45. Schmitt, *Back to Nature,* xvi–xvii. Schmitt quotes William Smythe's *City Homes on Country Lanes* (New York: Macmillan, 1922), 60.

46. Shi, *The Simple Life,* 196–97.

47. Earl Pomeroy, *In Search of the Golden West: The Tourist in Western America* (New York: Alfred A. Knopf, 1957), 143–45.

48. Cindy S. Aron, *Working at Play: A History of Vacations in the United States* (New York: Oxford University Press, 1999), 157.

49. Pomeroy, *Golden West,* 145.

50. Ibid., 166–73; Jared Farmer, "Legendary Peaks: Indian Plan Meets Alpine Play," paper delivered at the Western History Association annual conference, Colorado Springs, October 16–19, 2002.

51. John Ise, *Our National Park Policy: A Critical History* (Baltimore: Johns Hopkins University Press, 1961), 66.

52. Runte, *Yosemite,* 73.

53. Yosemite Park Commission, *Report of the Yosemite Park Commission,* pp. 8–9, as quoted in Runte, *Yosemite,* 75.

54. Ise, *Our National Park Policy,* 64, 86.

55. Ibid., 85–86; Jones, *John Muir and the Sierra Club,* 89–91; Runte, *Yosemite,* 76.

56. Jones, *John Muir and the Sierra Club,* 86. Although Jones's title does not reflect it, approximately one-half of his book is on the Hetch Hetchy fight. It is the best account in print, although Jones, as he freely admitted, looked at the issue only from the view of the Sierra Club and the papers that he perused. Jones's has the most complete description of the recession fight to reinstate Yosemite Valley to federal care.

57. John Muir, *The Mountains of California, The Writings of John Muir,* vol. 5 (Boston: Houghton Mifflin, 1916), 284–85.

2. THE IMPERIAL CITY AND WATER

1. Norris Hundley Jr., *The Great Thirst: Californians and Water, 1770s–1990s* (Berkeley and Los Angeles: University of California Press, 1992), 2–4. See also Gray Brechin, *Imperial San Francisco: Urban Power, Earthly Ruin* (Berkeley and Los Angeles: University of California Press, 1999), ch. 1, 13–70.

2. Hundley, *The Great Thirst*, 33–34.

3. Ibid., 50–57.

4. Ibid., 58–60.

5. Quoted in ibid., 65.

6. Alan Hynding, *From Frontier to Suburb: The Story of the San Mateo Peninsula* (San Mateo, Calif.: Star, 1982), 74.

7. Malcomb J. Rohrbough, *Days of Gold: The California Gold Rush and the American Nation* (Berkeley and Los Angeles: University of California Press, 1997), 203.

8. William S. Greever, *The Bonanza West: The Story of the Western Mining Rushes, 1848–1900* (Norman: University of Oklahoma Press, 1963), 52.

9. Rohrbough, *Days of Gold,* 246.

10. J. S. Holliday, *Rush for Riches: Gold Fever and the Making of California* (Berkeley and Los Angeles: University of California Press, 1999), 274–75.

11. Hundley, *The Great Thirst,* 76–77; Brechin, *Imperial San Francisco,* 50–51. Still the classic on the hydraulic mining controversy is Robert L. Kelley's *Gold vs. Grain: The Hydraulic Mining Controversy in California's Sacramento Valley—A Chapter in the Decline of Laissez-Faire* (Glendale, Calif.: Arthur H. Clark Co., 1959).

12. The dramatic story of mining versus agriculture is well told by Holliday, *Rush for Riches,* 267–304.

13. Brechin, *Imperial San Francisco,* 48.

14. Ibid., xxii–xxiv.

15. William Cronon, *Nature's Metropolis: Chicago and the Great West* (New York: W.W. Norton, 1991), 41–46.

16. Gerald T. Koeppel, *Water for Gotham* (New York: Oxford University Press, 2000), 5; also see J. G. Landels, *Engineering in the Ancient World* (Berkeley and Los Angeles: University of California Press, 1978).

17. Cronon, *Nature's Metropolis,* 43; Brechin, *Imperial San Francisco,* 73.

18. Holliday, *Rush for Riches,* 177. Holliday lists the dates of San Francisco's great fires as December 24, 1849; May 4, June 14, September 17, and December 14, 1850; and May 4 and June 22, 1851.

19. Ibid., 176. These fires, so destructive to property, were one of the primary reasons for the formation of the Vigilance Committee of 1851, which used either banishment or judicious hanging to restore order.

20. Most of the facts in these two paragraphs come from Marsden Manson, "The Struggle for Water in the Great Cities of the United States," *Association of Engineering Societies* 38 (March 1907), 103–24.

21. See Feral Egan, *Last Bonanza Kings: The Bourns of San Francisco* (Reno: University of Nevada Press, 1998), 106.

22. Laura Wood Roper, *A Biography of Frederick Law Olmsted* (Baltimore: Johns Hopkins University Press, 1973), 261.

23. Hynding, *From Frontier to Suburb,* 77.

24. Quoted in Brechin, *Imperial San Francisco,* 68.

25. J. R. McNeil, *Something New under the Sun: An Environmental History of the Twentieth-Century World* (New York: W. W. Norton, 2000), 150.

26. Hynding, *From Frontier to Suburb,* 76.

27. Ibid., 76.

28. John P. Young, *San Francisco: A History of the Pacific Coast Metropolis,* 2 vols. (San Francisco: S. J. Clarke Publishing Co., 1912), 1:586.

29. San Francisco's dominance by city bosses during the gilded age is well documented. See Alexander Callow Jr., "San Francisco's Blind Boss," *Pacific Historical Review* 25 (August 1956), 261–76; Walton Bean, *Boss Ruef's San Francisco* (Berkeley and Los Angeles: University of California Press, 1952; and William Bullough, *Blind Boss and His City: Christopher Augustine Buckley and Nineteenth-century San Francisco* (Berkeley and Los Angeles: University of California Press, 1979).

30. Robert W. Righter, "The Life and Public Career of Washington Bartlett" (M.A. thesis, History Department, San Jose State University, 1963), 55–59.

31. "San Francisco Will Win in the Hetch Hetchy Project," *The Grizzly Bear,* February 1909.

32. Brechin, *Imperial San Francisco,* 174–75.

33. Egan, *Last Bonanza Kings,* 174–76.

34. Hynding, *From Frontier to Suburb,* 110–11. The town that would become Hillsborough was first formed by William Ralston. As a development, it failed. When Ralston and his Bank of California also failed, William Sharon inherited the distressed assets of Ralston, which in turn fell into Newland's hands.

35. Douglas Strong, *Tahoe: An Environmental History* (Lincoln: University of Nebraska Press, 1984), 95–98.

36. Quoted in ibid., 96.

37. *San Francisco Examiner,* May 15, 1876, editorial.

38. Ibid.

39. *Oakland Daily Transcript,* May 5, 1875; *Oakland Daily News,* June 16, 1876.

40. Thomas P. Hughes, *Networks of Power: Electrification in Western Society, 1880–1930* (Baltimore: Johns Hopkins University Press, 1983), ch. 10 ("California White Coal"), 262–84, 265.

41. Strong, *Tahoe,* 98.

3. WATER, EARTHQUAKE, AND FIRE, 1901–1907

1. Walsh, O'Keefe, *Legacy of a Native Son,* 60–61.

2. See Jones, *John Muir and the Sierra Club,* 55–80.

3. See William Issel, Robert W. Cherny, *San Francisco, 1865–1932: Power, Politics, and Urban Development* (Berkeley and Los Angeles: University of California Press, 1986), 32–35.

4. Emily Wortis Leider, *California's Daughter: Gertrude Atherton* (Stanford: Stanford University Press, 1991), 204–5.

5. Walsh, O'Keefe, *Legacy of a Native Son,* 120.

6. James Phelan to Mabelle Gilman, May 8, 1906, in James Phelan Papers, 1880–1930, MSS C-B 800, box 1, Bancroft Library (hereafter abbreviated as BL).

7. Wanda Muir to John Muir, Adamada, Ariz., April 10, 1906, in John Muir Papers (microfilm), reel 16, 1906–07.

8. Kevin Starr, *Endangered Dreams: The Great Depression in California* (New York: Oxford University Press, 1996), 279.

9. Ibid., 121. Writing to the secretary of the interior in 1908, City Engineer Marsden Manson placed the date of entry as October 15, 1901. I accept the earlier date. See Marsden Manson Papers, C-B 416, carton 1:44, BL.

10. William Hammond Hall, "The Story of Hetch Hetchy," in William Hammond Hall Papers, 86/152, box 6, folder 6:18, BL. This lengthy and self-serving account of the Hetch Hetchy struggle tells the story from Hall's point of view.

11. Ibid., 4.

12. While Lippincott has received severe criticism for his actions in the Owens Valley, he has been practically invisible in the Hetch Hetchy Valley. Wherever he went, however, one finds evidence that he used his official office with either the U.S. Geological Survey or, after 1902, with the Bureau of Reclamation to feign neutrality while actually pushing the agenda of his client. For a severe criticism of Lippincott's behavior in the Owens River fight, see William L. Kahrl, *Water and Power: The Conflict over Los Angeles' Water Supply in the Owens Valley* (Berkeley and Los Angeles: University of California Press, 1982), 106–30.

13. *Reports on the Water Supplies of San Francisco, 1900–1908*, Board of Supervisors, 1908, 38–39. This 230-page compilation of important water documents can be found in the San Francisco History Room, San Francisco Public Library. It contains a number of the important documents from the time of C. E. Grunsky, City Engineer, 1900–1903, and Marsden Manson, Chief Engineer, 1908–1912. I believe Manson collected the documents and had them bound. It is a valuable source and, to my knowledge, is not easily—if at all—available elsewhere. I have used the materials quite extensively in this chapter.

14. John Muir to Robert Underwood Johnson, September 18, 1907, and September 23, 1907, in John Muir Papers, reel 16, 1907.

15. Hall, "The Story of Hetch Hetchy," 12, 30.

16. Quoted in ibid., 63.

17. Harvey Meyerson, *Nature's Army: When Soldiers Fought for Yosemite* (Lawrence: University Press of Kansas, 2001), 176–81. Throughout his book, Meyerson gives examples of the army's protection of Yosemite against despoilers, be they sheep men or San Francisco water seekers.

18. "A Brief in the Matter of Reservoir Rights of Way for a Domestic and Municipal Water Supply . . ." (July 27, 1907), 2, 4–5, in *Reports on the Water Supplies of San Francisco, 1900–1908,* Board of Supervisors, 1908, San Francisco Room, San Francisco Public Library.

19. *Reports on the Water Supplies* . . . , 23–24.

20. Secretary Hitchcock to President Theodore Roosevelt, February 20, 1905, in *Reports on the Water Supplies* . . . , 7, 21, 31.

21. Walsh, O'Keefe, *Legacy of a Native Son*, 79–80.

22. Ibid., 114.

23. "Report on Water Supply," by the Special Committee on Water Supply, dated October 8, 1906, 7, in SF City Engineer, 92/808 C, carton 3, folder "Data on Power," BL.

24. Ibid., 9.

25. Edwin Duryea, "The Facts about the Bay Cities Water Company's Water Supply for San Francisco," an address before the Commonwealth Club, March 13, 1907, 4–12 (copy in the Yosemite Archives, Yosemite National Park).

26. "Proposition Made by the Bay Cities Water Company to the Board of Supervisors," April 9, 1906, 16, attached to ibid.

27. See Charles M. Coleman, *P. G. and E. of California: The Centennial Story of Pacific Gas and Electric Company, 1852–1952* (New York: McGraw-Hill, 1952), 162–78.

28. See, for instance, Walsh, O'Keefe, *Legacy of a Native Son,* 114; Brechin, *Imperial San Francisco,* 107; David Lavender, *California: Land of New Beginnings* (New York: Harper and Row, 1972), 358.

29. Wanda Muir to John Muir, April 18, 1906, in John Muir Papers, reel 16, 1906–07.

30. "Report of the Sub-Committee on Water Supply and Fire Protection to the Committee on the Reconstruction of San Francisco," dated May 26, 1906, in *Reports on the Water Supplies of San Francisco,* Board of Supervisors, 1908.

31. Augustus Ward to Robert Underwood Johnson, July 21, 1908, in Robert Underwood Johnson Papers, C-B 385, box 2, Augustus Ward folder, BL.

32. A. H. Payson, "Letter from the President of the Spring Valley Water Company to the Special Committee of the Board of Supervisors on Water Supply," dated April 13, 1908, in *Reports on the Water Supplies of San Francisco,* Board of Supervisors, 1908.

33. *Congressional Record,* 63rd Cong., 1st sess., October 4, 1913, 6041.

34. Gifford Pinchot to Marsden Manson, May 28, 1906; Gifford Pinchot to Marsden Manson, November 15, 1906, both in *Reports on the Water Supply of San Francisco, 1900–1908,* Board of Supervisors, 1908.

35. Walsh, O'Keefe, *Legacy of a Native Son,* 125–26.

36. "Transcript of a Hearing before Honorable James R. Garfield in San Francisco, on the Evening of July 24, 1907," in *Reports on the Water Supply of San Francisco, 1900–1908,* Board of Supervisors, 1908.

37. John Muir to Theodore Roosevelt, September 7, 1907; Theodore Roosevelt to John Muir, September 16, 1907, both in John Muir Papers, reel 16, 1907.

38. John Sears, *Sacred Places: American Tourist Attractions in the Nineteenth Century* (New York: Oxford University Press, 1989), 185–89.

39. Ernest Morrison, *J. Horace McFarland: A Thorn for Beauty* (Harrisburg: Pennsylvania Historical and Museum Commission, 1995), 108–12.

40. Ibid., 116–20.

41. Wallace Stegner, *Beyond the Hundredth Meridian* (Boston: Houghton Mifflin, 1953), 242.

42. Morrison, *J. Horace McFarland,* 153.

43. The resolution in its entirety can be found in Jones, *John Muir and the Sierra Club,* 95–96. Major Harry Coupland Benson, the acting superintendent of Yosemite

under the War Department, also recommended a road to Hetch Hetchy in 1907, so that "a beautiful place would be opened to the general public." See 1907 Superintendent's Report, Yosemite Archives, Yosemite National Park.

44. Ibid., 97.

45. Hal Crimmel, "No Place for 'Little Children and Tender, Pulpy People': John Muir in Alaska," *Pacific Historical Review* 92 (Fall 2001), 172.

46. John Muir, *My First Summer in the Sierra* (Boston: Houghton Mifflin, [1911] 1972), 314.

47. Wolfe, *Son of the Wilderness,* 161.

48. John Muir to William Keith, October 7, 1907, in John Muir Papers, reel 16, 1907.

49. John Muir to Helen Muir, October 16, 1907; John Muir to William Colby, October 17, 1907, both in ibid.

50. John Muir to Theodore Lukens, November 4, 1907; John Muir to Charles Lummis, November 4, 1907, in ibid.

4. TWO VIEWS OF ONE VALLEY

1. "Sierra Club Beginnings," n.d., printed article by Ethel Olney Easton, with "Recollections of John Muir," typescript, no. 134, Regional Oral History Office, University of California.

2. John Muir to James Garfield, September 6, 1907, in John Muir Papers, reel 17, 1907.

3. See, for instance, the 1989 documentary film *The Wilderness Idea: John Muir, Gifford Pinchot and the First Great Battle for Wilderness,* which misses the essence of the struggle.

4. Char Miller, *Gifford Pinchot and the Making of Modern Environmentalism* (Washington, D.C.: Island Press, 2001), 126–27.

5. Gifford Pinchot, *Breaking New Ground* (Seattle: University of Washington Press, [1972] 1947), 103.

6. Wolfe, *Son of the Wilderness,* 275–76.

7. Miller, *Gifford Pinchot,* 122–23.

8. In both letters and documents between 1905 and 1908 Pinchot and Marsden Manson suggested that Hetch Hetchy should remain undeveloped until needed, for a period estimated anywhere from 30 to 100 years.

9. Robert Underwood Johnson to President Theodore Roosevelt, April 28, 1908, John Muir Papers, reel 17, 1908.

10. John Muir to Robert Underwood Johnson, March 11, 1908, John Muir Papers, reel 17, 1908; John Muir to Theodore Lukens, March 11, 1908, Theodore P. Lukens Papers, box 2, Correspondence, Muir folder, Huntington Library.

11. John Muir to President Theodore Roosevelt, April 21, 1908; President Roosevelt to John Muir, April 27, 1908, both in John Muir Papers, reel 17, 1908.

12. Marsden Manson to the Honorable Secretary of the Interior, May 7, 1908, Marsden Manson Papers, C-B 416, carton 1:44, BL.

13. "Decision of the Secretary of the Interior Department, Washington, D.C.,

Granting the City and County of San Francisco, Subject to Certain Conditions, Reservoir Sites and Rights of Way at Lake Eleanor and Hetch Hetchy Valley in Yosemite National Park," *Reports on the Water Supply of San Francisco, 1900–1908,* Board of Supervisors, 1908.

14. John Muir to Robert Underwood Johnson, May 14, 1908; John Muir to William Colby, May 15, 1908, in John Muir Papers, reel 17, 1908.

15. Jones, *John Muir and the Sierra Club,* 100.

16. "To the President of the United States and the Governors of the States assembled in Conference, from the Sierra Club Board of Directors," May 2, 1908, in John Muir Papers, reel 17, 1908.

17. "Hetch Hetchy Damming Scheme," memorandum from John Muir, President of the Sierra Club, received May 14, 1908, by J. Horace McFarland, President, American Civic Association; Robert Underwood Johnson to John Muir, May 23, 1908, both in ibid.

18. Quoted in Hans Huth, *Nature and the Americans: Three Centuries of Changing Attitudes* (Lincoln: University of Nebraska Press, 1972), 187.

19. William Colby to Robert Underwood Johnson, August 17, 1908, Robert Underwood Johnson Papers, C-B 385, box 2, Colby folder, BL.

20. See, for instance, "Mr. John Muir's Reply to a Letter Received from Hon. James R. Garfield . . ." [circa May 1908], John Muir Papers, reel 17, 1908; and "Let Everyone Help to Save the Famous Hetch Hetchy Valley . . . ," dated November 1909, copy in Hetch Hetchy Collection, DeGolyer Library, Southern Methodist University.

21. U.S. Congress, House of Representatives, Committee on the Public Lands, "San Francisco and the Hetch Hetchy Reservoir," Hearings on H.J. Resolution 184, 60th Cong., 1st Sess., December 16, 1908–January 12, 1909, 98.

22. John Muir to Robert Underwood Johnson, December 1, 1908, John Muir Papers, reel 17, 1908.

23. John Muir, J. N. LeConte, E. T. Parsons, and Wm. F. Badè, "To All Lovers of Nature and Scenery," December 21, 1908, John Muir Papers, Reel 17, 1908.

24. House, Committee on the Public Lands, "San Francisco and the Hetch Hetchy Reservoir," 116–243.

25. Ibid., 69, 91–4.

26. Ibid. A. P. Giannini was president of the Bank of Italy, which evolved into the Bank of America. Later, under his direction, the Bank of Italy purchased many of the city's Hetch Hetchy bonds when no other bank or investor would do so.

27. John Muir to William Colby, December 31, 1908, John Muir Papers, reel 17, 1908.

28. U.S. Congress, Senate, Committee on the Public Lands, "Hearings on the Hetch Hetchy Reservoir Site," Joint Resolution S.R. 123, 60th Cong., 1st Sess., February 10, 1909, 7, 24.

29. Ibid., 4–5, 30–33, 23.

30. Ibid., 18–21.

31. Ibid., 25–29.

32. *San Francisco Call,* February 11, 1909.

33. *Collier's Weekly,* reprinted in the *San Francisco Bulletin,* March 10, 1910.

34. Hall, "The Story of Hetch Hetchy," 10. This interesting 99-page typescript is located in the William Hammond Hall Papers, Bancroft Library. In this essay, undated but written about 1925, Hall wanted to tell his side of the Lake Eleanor controversy, which certainly had sullied his reputation in the city. He probably wished to publish this account but failed because it is lengthy, repetitive, and above all, self-serving. In his rationale for such a long account, Hall placed inordinate confidence in his own ability to tell the story and his faith that "it is only autobiography which can be both full and altogether truthful." The manuscript, although useful in many ways, is a refutation of his statement.

35. Hall, "The Story of Hetch Hetchy," 18–19.

36. Quoted in Brechin, *Imperial San Francisco,* 58.

37. Hall, "The Story of Hetch Hetchy," 71.

38. C. D. Marx and J. D. Galloway, "Report on Lake Eleanor Lands," August 28, 1909, in Charles D. Marx Papers, SC 161, box 3, folder 7, Stanford University Archives.

39. Hall came out of the affair a bitter man. He had a promising career as California's second state engineer, following Josiah Whitney. He then worked miracles in transforming San Francisco's barren sand hills into Golden Gate Park. He represented John Wesley Powell's irrigation survey in California. However, all these accomplishments were overshadowed by his manipulative behavior, and he became a *persona non grata* in the city he loved. He lived on in San Francisco much embittered as a testament that avarice can destroy an illustrious career. He died unnoticed in 1934, just a few weeks before the Hetch Hetchy water flowed to San Francisco. See Brechin, *Imperial San Francisco,* 81–84.

40. "The Hetch Hetchy Project: Its Progress, Prospects and Possibilities," January 1919, 2, in SF City Engineer, 92-808 C, carton "Writings on Hetch Hetchy," folder "January, 1919," BL.

41. Telegram from Phelan to Theodore Roosevelt, March 22, 1909, James D. Phelan Papers, C-B 800, box 2, folder "March, 1909," BL.

42. Richard Ballinger to Robert Underwood Johnson, June 11, 1909, John Muir Papers, reel 18, 1909.

43. Robert Underwood Johnson to John Muir, September 7, 1909; Frank Carpenter, Secretary to the President, to John Muir, September 6, 1909; John Muir to Robert Underwood Johnson, September 14, 1909, all in ibid.

44. *San Francisco Call,* October 10, 1909; *San Jose Herald,* October 10, 1909.

45. Muir to Colby, October 21, 1909; Muir to Katherine Hooker and Marian, October 20, 1909, in John Muir Papers, reel 18, 1909.

46. John Muir to Richard Ballinger, Secretary of the Interior, November 15, 1909, John Muir Papers, reel 18, 1909.

47. An original pamphlet is in the DeGolyer Library, Southern Methodist University. Few of the pamphlets have survived.

48. See Jones, *John Muir and the Sierra Club,* 97–99.

49. Ibid., 98–99.

50. Ibid., 112–13.

51. Quoted in ibid., 117.

52. Fox, *John Muir and His Legacy,* 119–20.

53. William Colby to Horace McFarland, September 13, 1909, in SC Members Papers, 71/295, carton 38, folder 18, BL.

54. Harriet Monroe, *A Poet's Life: Seventy Years in a Changing World* (New York: Macmillan, 1938), 218.

55. Harriet Monroe, "Camping above the Yosemite—A Summer Outing with the Sierra Club" *Putnam's Monthly* (1909), also published in the *Sierra Club Bulletin* 7 (June 1909): 85–98. Monroe wrote lovingly of "the glory of the wilderness" and appreciated the fact that she "had possessed it before its ways are made smooth for all the world." Yet she realized that the Hetch Hetchy wilderness would soon change. It would soon be "made smooth," and she did not begrudge that outcome.

Monroe may have taken some liberty with her description. Although no "spade" or plow had intruded in the valley, in 1908 there was at least one substantial building—probably three, depending on how one defines "hut."

56. Brief of Miss Harriet Monroe to Hon. Richard A. Ballinger, n.d., 3–5, Sierra Club Members Papers, 71/295, carton 38, folder 13, BL.

57. William Colby to Harriet Monroe, March 3, 1910, Sierra Club Members Papers, 71/295 C, carton 38, folder 21, BL.

58. Jones, *John Muir and the Sierra Club,* 103–4, 134.

59. Theodore Hittell, "Sierra Club Outing 1901," ed. Robert Righter, *Sierra Club Bulletin* 55 (August 1970), 18–22.

60. Carolyn Merchant, "Women and Conservation," 373–82, in *Major Problems in American Environmental History*, ed. Carolyn Merchant (Lexington, Mass.: D. C. Heath, 1993).

61. Samuel P. Hays, *Conservation and the Gospel of Efficiency* (Cambridge, Mass.: Harvard University Press, 1959), 193–94.

62. William Colby to Mrs. F. W. Gerald, G.F.W.C., n.d., circa July 1909; Jessie B. Gerald to Wm. Colby, September 11, 1909, in Sierra Club Members Papers, 71/295, carton 38, folder 18, BL.

63. Dora Knowlton Ranous to R. U. Johnson, undated, Robert Underwood Johnson Papers, C-B 385, box 5, Dora Knowlton Ranous folder, BL.

64. See Merchant, "Women and Conservation," 373.

65. E. T. Parsons to William Colby, September 7, 1909, Sierra Club Members Papers, 71/295, carton 38, folder 18, BL. Parsons and his wife, Marion, were in contact with Janet Richards.

66. Michael Cohen, *The History of the Sierra Club, 1892–1970* (San Francisco: Sierra Club Books, 1988), 64; Sierra Club, http://www.sierraclub.org/history/key_figures/"Marion Randall Parsons." Accessed by author on March 2, 2002.asp.

67. Marsden Manson, "Observations on the Denudation of Vegetation: A Suggested Remedy for California," *Sierra Club Bulletin*, 2 (June 1899), 295–311. Sounding much like George Perkins Marsh, Manson observed that "history and nature record no law more inflexible—no effect more certain—than that poverty and degradation follow upon the destruction of mountain forests."

68. Quoted in Smith, *Pacific Visions,* 177–78.

69. Nash, *Wilderness and the American Mind,* 167: Smith, *Pacific Visions,* 178. Both Nash and Smith attribute this quote to a letter from Manson to G.W. Woodruff, April 6, 1910, in the Manson Papers, BL. In perusing the Manson Papers, I missed it but have no cause to doubt that he wrote it.

70. *San Francisco Call,* reproduced in Jones, *John Muir and the Sierra Club,* 183.

71. Henry James, *The Bostonians* (New York: New American Library, 1984), 203.

72. Smith, *Pacific Visions,* 176. Smith provides the most insightful analysis of the genderization of the Hetch Hetchy conflict.

73. Marsden Manson, "A Statement of San Francisco's Side of the Hetch Hetchy Reservoir Matter, December 30, 1909," address to the Sierra Club, in *Pamphlets on Hetch Hetchy,* bound volume in the Bancroft Library, F869 S3.8 P25x.

74. Smith, *Pacific Visions,* 177. Smith gives a detailed account of the Fisher-Manson-McFarland 1911 horse trip to the Hetch Hetchy Valley in chapter 5 of this book.

75. Kendrick A. Clements, "Politics and the Park: San Francisco's Fight for Hetch Hetchy, 1908–1913," *Pacific Historical Review* 48 (May 1979), 200.

76. Donald Worster, "John Muir and the Roots of Environmentalism," in *The Wealth of Nature: Environmental History and the Ecological Imagination* (New York: Oxford University Press, 1993), 194; Catherine L. Albanese, *Nature Religion in America: From the Algonkian Indians to the New Age* (Chicago: University of Chicago Press, 1990), 101.

77. John Muir, "The Hetch Hetchy Valley," *Sierra Club Bulletin* 6 (January 1908), 211.

78. *San Francisco Call,* January 27, 1909; *San Francisco Bulletin,* February 28, 1910.

79. Beverly Hodghead, "The Hetch Hetchy Water Supply," *Transactions of the Commonwealth Club* 4, no. 6 (November 1909).

80. "Hill/Hopson Report made to George O. Smith, United States Geological Survey," December, 1909 (includes critical letters regarding San Francisco by both E. G. Hopson and Louis Hill) in *Reports on the Water Supply of San Francisco,* Board of Supervisors, 1908.

5. SAN FRANCISCO TO "SHOW CAUSE"

1. Horace McFarland, American Civic Association, to William Colby, Executive Director, Sierra Club, February 4, 1910, in Sierra Club Members Papers, 71/295 C, carton 38, folder 20, BL (hereafter Sierra Club Members Papers are abbreviated as SCMP.)

2. John Muir to Hon. Richard A. Ballinger, March 30, 1910, SCMP 71/295 C, carton 38, folder 21, BL.

3. Jones, *John Muir and the Sierra Club,* 123.

4. Richard Watrous, Secretary, American Civic Association, to William Colby, March 23, 1910, in SCMP 71/295 C, carton 38, folder 21, BL.

5. Fox, *John Muir and His Legacy,* 333, 333–57.

6. William Frederic Badè, ed., *The Life and Letters of John Muir,* 2 vols. (New York: Houghton Mifflin, 1923–1924): John Muir, *A Thousand Mile Walk to the Gulf,* ed.

William F. Badè (Boston: Houghton Mifflin, 1916). Badè also edited *Steep Trails* and *The Cruise of the Corwin*.

7. William Colby to William Badè, April 19, 1910, Badè to Colby, May 18, 1910, SCMP, 71/295 C, carton 38, folder 21, BL.

8. Wm. Badè to Wm. Colby, May 18, 1910, SCMP 71/295 C, carton 38, folder 21, BL.

9. Joseph N. LeConte should not be confused with his father, Joseph LeConte, who died in 1901. Both were active, charter members of the Sierra Club. The son was known as "Little Joe."

10. Charles Gilman Hyde to Wm. Colby, April 25, 1910; Hyde to Colby, May 6, 1910; Colby to Muir, May 16, 1910, in SCMP 71/295 C, carton 38, folder 21, BL.

11. Albert L. Hurtado, "Romancing the West in the Twentieth Century: The Politics of History in a Contested Region," *Western Historical Quarterly* 32 (Winter 2001), 418, 417–25.

12. William Colby to John Muir, May 16, 1910, SCMP 71/295 C, carton 38, folder 21, BL.

13. E. T. Parsons to Hon. Richard A. Ballinger, February 7, 1910, SCMP 71/295 C, carton 38, folder 20; Jones, *John Muir and the Sierra Club,* 125–26.

14. Jones, *John Muir and the Sierra Club,* 126.

15. Miller, *Gifford Pinchot,* 207.

16. Richardson, *Politics of Conservation,* 72–84; Miller, *Gifford Pinchot,* 208–9.

17. Richardson, *Politics of Conservation,* 136–37.

18. Stephen T. Mather to Wm. Colby, April 4, 1911, SCMP 71/295 C, carton 38, folder 23, BL. At this time Mather was the owner of Thorkildsen-Mather Company of Chicago, specializing in borax and soda ash. Colby and Mather were already good friends, and Mather comments on the pleasurable afternoon he spent at Colby and his wife's home.

19. Copy of letter from Percy Long, City Attorney to Walter L. Fisher, Sec. of the Interior, January 20, 1912, SCMP 71/295 C, carton 38, folder 24, BL.

20. Freeman's salary was a source of speculation. Two sources mention Freeman's consulting fee at $400 an *hour.* But the sources are engineers who did not like Freeman and were probably jealous of his success. $400 a *day* would still be a princely rate. S. L. Foster to Wm. Colby, October 2, 1912, SCMP, 71/295 C, carton 38, folder 25, BL. Freeman's actual wages included a $2,500 yearly retainer, $200 a day for work in the field, and $100 a day for work at his home office. These conditions were spelled out in a letter from Freeman to James Rolph, mayor, dated September 23, 1912, affixed to the inside back page of Marsden Manson's annotated copy of the Freeman *Report,* found in Marsden Manson Papers, C-B 416, carton 2:4, BL.

21. Sec. Fisher to Mayor of San Francisco, May 20, 1912, SCMP 71/295 C, carton 38, folder 24, BL.

22. Walter A. Fisher to Mayor James Rolph, Jr., May 28, 1912; J. Horace McFarland to Wm. Colby, June 1, 1912; Allen Chamberlain to Wm. Colby, June 3, 1912, SCMP, 71/295 C, carton 38, folder 24, BL.

23. *The Hetch Hetchy Water Supply for San Francisco, 1912* report by John R. Freeman (San Francisco: Rincon Publishing Company for the Board of Supervisors, 1912). I was so impressed with the size and weight of the report that I asked Southern

Methodist University DeGolyer Library former director David Farmer if he might weigh the volume. He took it to the adjacent post office and on his return announced: five pounds, eight ounces.

24. Ibid., 43, 46.

25. Ibid., 6, 10.

26. Ibid., 147–48.

27. Ibid., 149–51.

28. Ibid., 138–39.

29. Freeman's estimate of the Bay Area population was reasonably close. The 2000 Census records show that San Francisco, San Mateo, Santa Clara, and Alameda—the counties that border the San Francisco Bay—have a population of 4,623,166. Marin County was not part of Freeman's calculations.

30. Ibid., 161–401.

31. Donald C. Jackson, *Building the Ultimate Dam: John S. Eastwood and the Control of Water in the West* (Lawrence: University Press of Kansas, 1995), 111.

32. Robert Underwood Johnson to Wm. Colby, September 11, 1912; J. Horace McFarland to Wm. Colby, September 11, 1912; Harriet Monroe to Wm. Colby, September 17, 1912: Wm. Colby to J. Horace McFarland, October 16, 1912, SCMP 71/295 C, carton 38, folder 25, BL.

33. Starr, *Endangered Dreams,* 282. Starr relates the story of Hetch Hetchy in a chapter titled "Valley in Discord."

34. Colby Brief, 7–8, SCMP 71/295 C, carton 38, folder 13, BL. Jones, *John Muir and the Sierra Club*, quotes a large part of the brief (135–45), but I have chosen to use the copy in the above collection.

35. Ibid.

36. George Wharton James, *Winter Sports at Huntington Lake Lodge in the High Sierras* (Pasadena, Calif.: Radiant Life Press, 1916), tells the story of this first winter carnival. By the 1930s the popularity of winter sports led to the establishment of Badger Pass in Yosemite National Park, the first winter sports center in California. Obviously, the Colby Brief presaged the coming economic importance of winter sports.

37. Colby Brief, 26–29. The proponents of the dam suggested that once the Hetch Hetchy "soft water" was available, women would love it because in washing clothes they would use only a half or a third of the soap they were then using. Hence Colby's mention of soap.

38. The road recommendation came as part of a first formal statement opposing San Francisco's plans drafted by Muir, Colby, "Little Joe" LeConte, Wm. Badè, and Edward Taylor Parsons, quoted in Jones, *John Muir and the Sierra Club,* 95–97. Also see Cohen, *History of the Sierra Club,* 25.

39. John Muir, *Our National Parks* (Boston: Houghton Mifflin, 1916), 3–4. In 1912 neither Muir nor Colby anticipated the number of automobiles that were mass-manufactured in the 1920s and that descended on Yosemite National Park.

40. Issel, Cherny, *San Francisco, 1865–1932,* 161–62.

41. I am in debt to Kevin Starr, *Endangered Dreams,* 284, for the description of Mayor Rolph.

42. Michael O'Shaughnessy was the first engineer to consider seriously the possi-

bility of a Golden Gate Bridge connecting San Francisco to Marin County and the northern regions. Had the Hetch Hetchy idea not moved forward, it is very possible that O'Shaughnessy might have devoted his engineering talents to designing and advocating the great bridge. See Starr, *Endangered Dreams,* 330–31.

43. M. M. O'Shaughnessy, "Irrigation Works in the Hawaiian Islands," *Engineering News* 61 (April 15, 1909), 399–403. The Olokele canal is still maintained today on the Big Island, and as an outdoor adventure, one can take a two-and-a-half-mile kayak trip.

44. *Los Angeles Times,* August 30, 1908.

45. Michael M. O'Shaughnessy, *Hetch Hetchy: Its Origins and History* (San Francisco: Privately printed by the Recorder Printing and Publishing Company, 1934), 13. O'Shaughnessy's recollections contain a number of verbatim letters as well as his interpretation of the fight.

46. Rose Wilder Lane, "The Building of Hetch Hetchy," serial account running in the *San Francisco Bulletin,* October 14–November 14, 1916. Composed from interviews with O'Shaughnessy.

47. O'Shaughnessy, *Hetch Hetchy,* 22–23; see also "Site Inspection, September, 1912," 18 pp., in SF City Engineer, 92/808 C, carton 3, folder "Reports, 1912", BL.

48. J. Horace McFarland to Wm. Colby, September 27, 1911, SCMP 71/295 C, carton 38, folder 23, BL.

49. Ibid.

50. Ibid.

51. Ibid.

52. Robert Marshall, U.S. Geological Survey, to Wm. Colby, September 29, 1911, in ibid.

53. J. D. Dockweiler, *Report on Sources of Water Supply, East Region of San Francisco Bay* (509-page typescript, prepared for the cities of Oakland and Berkeley, but jointly with San Francisco at the request of Percy V. Long, City Attorney), copy in the San Francisco History Room, San Francisco Public Library.

54. Hiram M. Chittenden, ed., *The Water Supply of San Francisco* (Spring Valley Water Company, 1912), 506 pages.

55. J. Horace McFarland to Wm. Colby, December 28, 1912, SCMP 71/295 C, carton 38, folder 25, BL.

56. Telegrams to Wm. Colby, Alden Sampson, November 25; McFarland, November 26; Badè, November 27; Sampson, November 29, November 30, in SCMP 71/295 C, carton 38, folder 25, BL.

57. J. Horace McFarland to Stephen Mather, December 2, 1912, in ibid.

58. See H. Duane Hampton, *How the U.S. Army Saved Our National Parks* (Bloomington: University of Indiana Press, 1971), and the more recent Meyerson, *Nature's Army* (2001).

59. Ben R. Martin, "The Hetch Hetchy Controversy: The Value of Nature in a Technological Society," Ph.D. dissertation, Department of Philosophy and History of Ideas, Brandeis University, 1981, 302; Carroll Purcell, *The Machine in America* (Baltimore: Johns Hopkins University Press), 140; Wm. Colby to Allen Chamberlain, January 12, 1912, SCMP 71/295 C, carton 34, folder 23, BL.

60. Smith, *Pacific Visions*, 180.

61. U.S. Senate, Committee on the Public Lands, "Hearings on Hetch Hetchy Reservoir Site," Joint Resolution S.R. 123, February 10, 1909, 9.

62. Clements, "Politics and the Park," 203.

63. *Hetch Hetchy. Report of Advisory Board of Army Engineers to the Secretary of the Interior* (Washington, D.C.: Government Printing Office, 1913), 14–15.

64. In 1920 O'Shaughnessy reported that Taylor, in 1920 a general, was "very much impressed" and "a sincere believer" in the development undertaken by the city. In 1923 O'Shaughnessy introduced Biddle to the city supervisors as a man responsible for the project. See SF City Engineer, 92/808 C, carton 2, folder "Reports, 1920," and carton 4, folder "Clippings, Jan.–Aug. 1923," BL.

65. Martin, "The Hetch Hetchy Controversy," 444.

66. Secretary Walter Fisher to Mayor James Rolph, March 1, 1913, quoted in Jones, *John Muir and the Sierra Club,* 151.

6. CONGRESS DECIDES

1. Horace McFarland to Wm. Colby, December 28, 1912, SCMP 71/295, carton 38, folder 25, BL.

2. Richard Watrous to Wm. Colby, March 10, 1913, SCMP 71/295, Carton 38, folder 26, BL. After the final defeat, Horace McFarland singled out Lane's appointment as the most important factor. See Horace McFarland to Wm. Colby, December 20, 1913, SCMP, 71/295, carton 38, folder 27, BL.

3. Horace M. Albright, Marian Albright Schenck, *Creating the National Park Service: The Missing Years* (Norman: University of Oklahoma Press, 1999), 20.

4. Arthur Link, ed., *The Papers of Woodrow Wilson*, 79 vols. (Princeton: Princeton University Press, 1979) 5:25, 558–59.

5. Ibid., 5:27, 126, 130–31. Also see Arthur Link, *Wilson and the New Freedom* (Princeton: Princeton University Press, 1956), 18–19. Contrary to House's assessment, Norman Hapgood did not represent the "extreme view" of Eastern conservationists. In 1913 he would endorse the San Francisco plan to invade Hetch Hetchy, much to the chagrin of Sierra Club members, such as Harold Bradley, a young man who would eventually assume a leadership role.

6. Lane, "The Building of Hetch Hetchy." The journalist interviewed O'Shaughnessy extensively; thus this story is probably true in general, but with some journalistic liberties.

7. Clements, "Politics and the Park," 206–7.

8. O'Shaughnessy, *Hetch Hetchy,* 46.

9. Pinchot's and Newell's and Phelan's testimony in U.S. Congress, House of Representatives, Committee on Public Lands, "Hetch Hetchy Grant to San Francisco" by Mr. Raker, Rep. no. 41, 63rd Cong., 1st sess., to accompany H.R. 7207, 25–31; also in *Congressional Record,* 63rd Cong., 1st sess., vol. 50 (August 29, 1913), 3907–9. Also see Jones, *John Muir and the Sierra Club,* 157.

10. Clements, "Politics and the Park," 209–11.

11. Ibid., 211–12, 3896.

12. *Congressional Record,* 63rd Cong., 1st sess., vol. 50 (August 29, 1913), 3972–73.

13. Ibid., 3974–75.

14. Ibid. (September 3, 1913), 4151; Thomson quote from ibid. (August 30, 1913), 3977.

15. O'Shaughnessy, *Hetch Hetchy,* 47.

16. See Hal Rothman, *Preserving Different Paths: The American National Monuments* (Chicago: University of Illinois Press, 1989), 61–64, for a more complete story of the politics surrounding Kent and the Muir Woods gift.

17. William Kent to John Muir, January 16, 1908; Muir to Kent, January 14, 1908, in John Muir Papers, reel 17, 1908.

18. Anne F. Hyde, "William Kent: The Puzzle of Progressive Conservationists," in *California Progressivism Revisited,* ed. William Deverell and Tom Sitton (Berkeley and Los Angeles: University of California Press, 1994), 34–53, passim.

19. *Congressional Record,* 63rd Cong., 1st sess., vol. 50, (August 29, 1913), 3963.

20. Clements, "Politics and the Park," 208, n. 66.

21. Quote from Wolfe, *Son of the Wilderness,* 337.

22. See E. T. Parsons to Wm. Colby, August 1, 1913, in Jones, *John Muir and the Sierra Club,* 159.

23. Robert Underwood Johnson, "The Hetch Hetchy scheme; why it should not be rushed through the extra session; an open letter to the American people," Library of Congress "An American Time Capsule: Three Centuries of Broadsides and Other Printed Ephemera," http://memory.loc.gov/cgi-bin/ampage. Digital ID http://hdl. loc,gov.vbc/vbpe 130080a Date Accessed, March 15, 2002.

24. William Howard Taft to Robert Underwood Johnson, October 30, 1913, in Robert Underwood Johnson Papers, C-D 385, "Taft" folder, BL.

25. Albert Bushnell Hart to Robert Underwood Johnson, November 10, 1913, in ibid., "Hart" folder, BL.

26. *Congressional Record,* 63rd Cong., 2nd sess., vol. 51 (November 25, 1913), 6012; Ashurst material from ibid., (December 5, 1913), 234–35. Senator Ashurst asked that some of the letters—about 20—be inserted into the record, but they all reflected his support of the Raker bill. Reed quote from ibid. (December 6, 1913), 362; Albright, Schenck, *Creating the National Park Service,* 20.

27. Fred L. Israel, *Nevada's Key Pittman* (Lincoln: University of Nebraska Press, 1963), 31.

28. "Speech of Hon. John D. Works of California in the Senate of the U.S., December 2–3, 1913," Washington, D.C., 6 (copy in the San Francisco History Room, San Francisco Public Library).

29. I am in debt to Richard Lowitt's article, "Hetch Hetchy Phase II: The Senate Debate," in *Frontier and Region: Essays in Honor of Martin Ridge,* ed. Robert C. Ritchie and Paul Andrew Hutton (Albuquerque: University of New Mexico Press, 1997), 109–20.

30. *Congressional Record,* 63rd Cong., 2nd sess., vol. 51 (December 5, 1913), 290–97.

31. Ibid. (December 4, 1913), 199–200. The next day, December 5, Senator Bacon chastised his fellow senators for absenting themselves the night before when Senator

Gronna had the floor, particularly since Gronna was "making an interesting and logical argument on the question" (243).

32. Ibid. (December 5, 1913), 273–74.

33. Ibid., 274.

34. Ibid. (December 6, 1913), 233–34.

35. Ibid. (December 5, 1913), 238; Lowitt, "Hetch Hetchy Phase II," 115–17.

36. *Congressional Record,* 63rd Cong., 2nd sess., vol. 51 (December 6, 1913), 344, 347.

37. Ibid., 382, 385–86.

38. O'Shaughnessy, *Hetch Hetchy,* 52.

39. *Congressional Record,* 63rd Cong., 2nd sess., vol. 51 (December 19, 1913), 1189.

40. Perhaps out of frustration, Muir wrote Robert Underwood Johnson a letter critical of the efforts of his West Coast leaders. In truth, they were tired and without financial support; they might be excused for letting up a bit, particularly when the hopelessness of their situation became clear. Letter quoted in Jones, *John Muir and the Sierra Club,* 160.

41. Clements, "Politics and the Park," 214.

42. The Senate did at one point discuss the possible financial loss to San Francisco regarding land and water rights purchases if the Raker bill failed of passage. The consensus was that the federal government would make it right with the city. Considering the nominal purchase price of land in Hetch Hetchy Valley, no doubt one of the first "inholding" purchases of the National Park Service (1916) would have been the meadowlands of the valley.

In regard to Wilson and Lane, I must again reiterate that there is no evidence that Wilson knew anything about Hetch Hetchy. And, it appears, Lane had not met Wilson prior to his appointment. Neither is there written evidence that Colonel House and Lane ever discussed the Hetch Hetchy issue. One can, however, make that dangerous leap of faith that the Colonel and Lane had discussed Hetch Hetchy in conjunction with their conversation on conservation policy.

43. Jones, *John Muir and the Sierra Club,* 166–67.

44. Quoted in ibid., 168.

45. Quoted in Wolfe, *Son of the Wilderness,* 341.

46. John Muir to Helen [Muir], January 12, 1914, John Muir Papers, reel 22, 1914.

47. Ibid.

48. Wolfe, *Son of the Wilderness,* 344.

7. TO BUILD A DAM

1. O'Shaughnessy, *Hetch Hetchy,* 47.

2. Ed L. Head to O'Shaughnessy, n.d.; Congressman John Raker to M. M. O'Shaughnessy, April 22, 1914, in SF City Engineer, 92/808 C, carton "Correspondence Apr. 1912–Aug. 1914," folder Dec. 1913, BL.

3. J. Horace McFarland to M. M. O'Shaughnessy, August 4, 1914, quoted in Morrison, *J. Horace McFarland,* 171–72. I don't wish to suggest that McFarland abandoned the national parks. He was instrumental in writing and lobbying for the National Parks Act of 1916.

4. M. M. O'Shaughnessy, "Hetch Hetchy Water Supply of the City and County of San Francisco, California," January, 1917, 4, in SF City Engineer, 92/808 C, box 1, folder "Reports 1917," BL.

5. "Discussion of a Proposed Plan for disposing of the Hetch Hetchy Water Supply Bonds of 1910," typescript, SF City Engineer, 92/808 C, carton 1, folder "Reports 1915," BL.

6. M. M. O'Shaughnessy, "Tentative Program of Construction, and Preliminary Estimate of Cost," May 7, 1915, ibid.

7. Ted Wurm, *Hetch Hetchy and Its Dam Railroad* (Berkeley: Howell-North Books, 1973), 28. By its title and at first glance, Ted Wurm's book seems written for railway buffs. However, because Wurm became more interested in the Hetch Hetchy system than in the railroad, there is a great amount of information on the building of the dam and the aqueduct system. Although Wurm is an unabashed apologist for the city, his oral history work and examination of esoteric materials provides details and anecdotes unavailable elsewhere.

8. "The History and Economic Aspect to Date of the San Francisco–Hetch Hetchy Water Supply," typescript of 30-page report by E. E. Schmitz, 10, 21, in SF City Engineer, 92/808 C, box 1, folder "Reports 1919," BL.

9. Ibid. Also see Rudolph W. Van Norden, "The Present Status of Hetch Hetchy," *Journal of Electricity* 41 (November 15, 1918), 438–43.

10. Wurm, *Hetch Hetchy,* 78–79.

11. Ibid. In his book on Hetch Hetchy, written about 1930, O'Shaughnessy states that the 880 acres on the reservoir bottom were "cleared for $50,000." But I prefer to use figures from his report at the time of the event.

12. Ise, *Our National Park Policy,* 194–95.

13. Quoted in Wurm, *Hetch Hetchy,* 63.

14. SF City Engineer, 92/808 C, box 1, folder "Reports 1916," BL

15. Wurm, *Hetch Hetchy,* 151.

16. Joseph E. Stevens, *Hoover Dam: An American Adventure* (Norman: University of Oklahoma Press, 1988), 3–5, 35–36.

17. Ibid. For the Hoover Dam the Wattis brothers led the charge that was instrumental in convincing such leaders as Warren Bechtel and Henry Kaiser to form the Six Companies, a consortium responsible for building the monumental dam that tamed the Colorado River.

18. O'Shaughnessy, *Hetch Hetchy,* 52.

19. This claim was justified, although the Hoover and Shasta dams, as well as the Chambon Dam in France, would soon claim the distinction. See William Creager, Joel D. Justin, Julian Hinds, *Engineering for Dams: Volume 2, Concrete Dams* (London: John Wiley and Sons, 1951), chart, 350–55.

20. "Contract 61 for the Construction of the Hetch Hetchy Dam and Appurtenant Work," in SF City Engineer, 92/808 C, carton 2, folder "Specifications 1919," BL.

21. Wurm, *Hetch Hetchy*, 86.

22. M. M. O'Shaughnessy, "Speech before the Commonwealth Club," November

19, 1919, in SF City Engineer, 92/808 C, carton "Writings Hetch Hetchy," folder "Nov. 19, 1919, speech," BL.

23. Wurm, *Hetch Hetchy,* 87, 285.

24. Ibid., 87–88.

25. Elford Eddy, "The Building of Hetch Hetchy," *San Francisco Call,* December 19, 1921–February 11, 1922.

26. William Hamilton, "Human Engineering," *Co-operation: The Journal of Efficiency,* October 1921, 12; Fred D. MacBeth, Attorney, to M. M. O'Shaughnessy, June 30, 1922, SF City Engineer, 92/808 C, box 1, folder "Corr. 1922," BL.

27. Hamilton, "Human Engineering," 20.

28. Story in the *San Francisco Chronicle,* October 27, 1934.

29. Strike leaflet, in Central Files, Hetch Hetchy General, 1917–1922, 660-05.41, Yosemite Archives, Yosemite National Park (hereafter abbreviated as YNPA).

30. SF City Engineer, 92/808 C, carton 2, folder "Reports 1922," BL. It is impossible to know exactly what Mr. Eckart meant when he referred to the "I.W.W. strike." But more than likely his reference was related to the mass arrests of IWW leaders by the Justice Department in 1917. In California about 20 members were arrested in September in Fresno for an alleged assassination attempt on California governor-elect William D. Stephens. They were held in jail for over a year. In January 1918 54 Wobblies were jailed in Sacramento. They remained incarcerated until finally, in December, the Justice Department brought them to trial. All were found guilty and sentenced to prison terms of one to ten years, but "they remained true to their IWW faith." No doubt some of these organizers and members, recently released from prison, influenced the Hetch Hetchy workforce. See Melvin Dubofsky, *We Shall Be All: A History of the IWW* (Chicago: Quadrangle Books, 1969), 438–41.

31. C. R. Rankin to W. B. Lewis, Supt., Yosemite National Park, October 16, 1922, Central Files, Hetch Hetchy General, 1917–1922, 660.05-41, YNPA.

32. SF City Engineer, 92/808 C, carton 2, folder "Memoranda 1921," BL; Starr, *Endangered Dreams,* 29.

33. Wurm, *Hetch Hetchy,* 136.

34. Ibid., 135–36.

35. Ibid., 136.

36. SF City Engineers, 92/808 C, carton 2, folder "Reports 1922," BL.

37. Ibid., carton 1, folder "1929–1930 Corr.," BL.

38. Speech to the Down Town Association, May 25, 1925, in SF City Engineers, 92/808 C, carton "Writings, Hetch Hetchy," folder "May 28, 1925," BL.

39. Fresno *Republican,* July 8, 1923.

40. "O'Shaughnessy Dam Dedication," *Municipal Record* 16 (July 19, 1923), in SF City Engineer, 92/808 C, carton 4, folder "O'Shaughnessy Dam Dedication scrapbook," BL.

41. Quoted in Jackson, *Building the Ultimate Dam,* 6.

42. Cronon, *Nature's Metropolis,* xvii, 56–57.

43. Beverly Hodghead to O'Shaughnessy, March 29, 1923, SF City Engineer, 920808 C, carton 4, folder "O'Shaughnessy Dam Dedication scrapbook," BL.

44. Clifford M. Holland to O'Shaughnessy, July 19, 1923; Milt Clark to O'Shaughnessy, n.d., in ibid.

45. Wurm, *Hetch Hetchy,* 187, 191. The electrical production and distribution system is covered in greater detail in chapter 8.

46. M. M. O'Shaughnessy, "Construction Progress of the Hetch Hetchy Water Supply of San Francisco, October, 1921," in SF City Engineers, 92/808 C, carton 2, folder "Reports 1922." BL.

47. Brechin, *Imperial San Francisco,* 72–73; see also Gray Brechin, "Water Rites: San Francisco's Water Temples Celebrate Classical Civilization," *Almost History*, August 1989, 14–17. My wife and I visited the Sunol Temple in the spring of 2002. Though the structure was severely damaged by the 1996 earthquake, the city of San Francisco has rebuilt it. According to Rob Cyr, a SFMUD employee for 20 years, the temple used to be a gathering palace for picnickers and hikers, but now, while it is open to the public, visitors are not encouraged to stay.

48. "Alameda County's Interest in Hetch Hetchy Grant," speech of Hon. Joseph R. Knowland of California in the House of Representatives, August 29, 1913, copy in the Huntington Library, San Marino, California.

49. M. M. O'Shaughnessy, "The Hetch Hetchy Project," *The Star*, June 1, 1920, 16–18, in SF City Engineer, 92/808 C, carton "Writings, Hetch Hetchy," folder "June 1, 1920," BL.

50. J. H. Kimball, EBMUD, to O'Shaughnessy, July 27, 1923; O'Shaughnessy to Kimball, July 28, 1923; Marston Campbell, EBMUD, to Mayor James Rolfe, February 15, 1924, in SF City Engineer, 92/808 C, Box 1, folder "Corres. 1923" and folder "Corres. 1924," respectively, BL.

51. O'Shaughnessy to John Freeman, April 1, 1924, Freeman to O'Shaughnessy, April 15, 1924, in ibid., folder "Corres. 1924."

52. M. M. O'Shaughnessy, "The Water Problem of the Bay Region," March 10, 1924, ibid., carton "Writings, Hetch Hetchy," folder "March 10, 1924," BL.

53. "The $39,000,000 Water Project of the East Bay Cities," *Engineering News-Record* 93 (December 11, 1924), 961.

54. "East Bay Hetch Hetchy League Resolution," July 26, 1924 in SF City Engineer, 92/808 C, carton "Writings, Hetch Hetchy, folder, 1924; "The $39,000,000 Water Project."

55. "The $39,000,000 Water Project," 960.

56. Some facts obtained from East Bay Municipal Utility District, http://www.ebmud.com/about_ebmud/overview/district_history.

57. Quoted in Brechin, *Imperial San Francisco,* 270.

58. "Transcript of meeting of mayor, board of supervisors, and the Advisory Water Committee, June 19, 1924," 11–12, in SF City Engineer, 92/808 C, carton 2, folder "Meeting, June 19, 1924," BL.

59. *San Francisco Bulletin*, November 14, 1923.

60. Wurm, *Hetch Hetchy,* 221.

61. Ibid., 222.

62. Egan, *Last Bonanza Kings,* 242–46. Bourn's FiLoLi estate, named for his motto of "Fight, Love, Live," is now managed for the public by the National Historic Trust.

63. SF City Engineer, 92/809 C, carton 3, folder "Site Inspection, March 1932," BL.

64. Ruth H. Willard, "Pulgas Water Temple," in Carol Green Wilson, ed., *Sacred Places of San Francisco* (San Francisco: Presidio Press, 1985); also in Book Club of California, *Newsletter* 52, no. 4 (n.d.).

65. *San Francisco Call,* October 12, 1934.

66. *San Francisco Chronicle*, October 29, 1934; "Celebration of the First Delivery of Hetch Hetchy Water to San Francisco's Crystal Springs Lake, San Mateo County," program for the October 28, 1934, opening of the Pulgas Temple, in BL; Harold L. Ickes, *The Secret Diary of Harold L. Ickes,* vol. 1, *The First Thousand Days, 1933–1936* (New York: Simon and Schuster, 1953), 214.

67. "Celebration of the First Delivery of Hetch Hetchy Water."

68. Hundley, *The Great Thirst*, 275.

69. Leslie W. Stocker, "Some Engineering Features of the Enlargement of the O'Shaughnessy Dam," *Journal of the American Water Works Association* 27 (August 1935), 986–92.

70. Ibid.

71. See correspondence between Harold Ickes to Elwood Meade, December 1934, in Department of the Interior, Office of the Secretary, Cent. Classified Files, 1907–1936, files 12, 13, pt. 6, National Archives, College Park (hereafter abbreviated as NACP).

8. THE POWER CONTROVERSY

1. Pinchot's views are best expressed in Gifford Pinchot, *The Power Monopoly: Its Make-up and Its Menace* (Milford, Pa., 1928), a privately printed tract in which he accused "investor-owned" companies in Pennsylvania of "ruthless exploitation, uninterrupted and unrestrained by anything approaching effective Government intervention and control." He felt the same animosity toward the Pacific Gas and Electric Company.

2. Richard F. Hirsh, *Power Loss: The Origins of Deregulation and Restructuring in the American Electric Utility System* (Cambridge, Mass.; MIT Press, 1999), 13–14.

3. Hughes, *Networks of Power,* 266.

4. *Congressional Record,* Senate, 63rd Cong., 1st sess., vol. 50 (December 6, 1913), 347.

5. Judson King, *The Conservation Fight: From Theodore Roosevelt to the Tennessee Valley Authority* (Washington, D.C.: Public Affairs Press, 1959), 42.

6. H.R. 7207, 63rd Cong., 1st sess., passed the House of Representatives on September 3, 1913. J. Horace McFarland and others credited Kent with adding section 6. See Morrison, *J. Horace McFarland,* 170.

7. George Norris, *Fighting Liberal: The Autobiography of George W. Norris* (New York: Macmillan, 1945), 162.

8. SF City Engineer, 92/808 C, box 1, folder "Reports 1916," 11, BL.

9. M. M. O'Shaughnessy to Herbert Hoover, Secretary of Commerce, November 8, 1923, in Dept. of the Interior, Office of the Secretary, Cent. Classified Files,

1907–1937, file "April 1923 to December 1923," box 2000, National Archives, College Park (hereafter abbreviated as NACP).

10. *San Francisco Examiner,* September 9, 1923.

11. "An Open Letter to Our City Engineer," *San Francisco Examiner,* December 10, 1923.

12. "Report on open meeting at the St. Francis Hotel sponsored by the Commonwealth Club," *San Francisco Daily News,* November 10, 1923.

13. Issel, Cherny, *San Francisco, 1865–1932,* 182–83.

14. Ibid., 184.

15. Agreement between the City of San Francisco and the Pacific Gas and Electric Company, signed July 1, 1925, copy in Records of U.S. Attorneys and Marshalls, Northern Department of California, San Francisco, Civil Cases, 1899–1950, 4173, RG 118, box 6, National Archives, San Bruno.

16. Telegrams from Mayor James Rolph to President Calvin Coolidge, June 26, 1925; Gifford Pinchot to Secretary Hubert Work, June 27, 1925; William Kent to Secretary Hubert Work, June 26, 1925, in Dept. of the Interior, Office of the Secretary, Cent. Classified Files, 1907–1936, RG48, box 2002, loose folder, NACP. Also see Issel, Cherny, *San Francisco, 1865–1932,* 184.

17. *Washington Daily News,* August 11, 1925.

18. *Los Angeles Record,* August 24, 1925.

19. Mayor James Rolph to Honorable Ray Lyman Wilbur, June 19, 1930, in Dept. of the Interior, Office of the Secretary, Cent. Classified File, 1907–1936, RG48, box 2002, letters folder, NACP. Also see "Statistics from Hetch Hetchy Water Supply & Power Project," Special Report by the Dept. of the Interior upon Financial History for Raker Act Hearings, Washington, D.C., April 8, 1935, in Dept. of the Interior, Office of the Secretary, Cent. Classified Files, 1907–1936, RG48, box 2004, NACP.

20. "Memorandum by Lewis F. Byington, President of the Public Utilities Commission of the City and County of San Francisco," 9–10, before the Secretary of the Interior, dated May 6, 1935, in Dept. of the Interior, Office of the Secretary, Cent. Classified Files, 1907–1936, RG 48, box 2000, files 12, 13, pt. 7, NACP.

21. Ibid., 1.

22. Hirsh, *Power Loss,* 14–15.

23. "Memorandum by Lewis F. Byington," 10–11.

24. Memorandum for the Secretary, from Arthur Demaray, Acting Director, National Park Service, June 27, 1933, in Dept. of the Interior, Office of the Secretary, Cent. Classified Files, 1907–1936, RG 48, box 2000, files 12, 13, pt. 6, NACP.

25. "In re the contract between San Francisco and the Pacific Gas and Electric Company in relationship to the Raker Act, August 24, 1935," by Harold L. Ickes, Secretary of the Interior, in ibid., pt. 7.

26. Advertisement against Amendment no. 1, *San Francisco Chronicle,* April 30, 1935.

27. Perhaps the most inflammatory letters came from Charles Pyral, who identified himself as an "Illumination Engineer" and the head of the "Municipal Ownership League of California." Pyral recounted stories of how the PG&E paid for supervisors to party at "booze fests" at an "Italian boot-legging joint" where supervisors

were wooed by PG&E executives. Such activities were part of the "power trusts'" stranglehold on San Francisco and the Bay Area. Pyral named the perpetrators of such activities, and although Ickes was skeptical, he did thank Pyral for his efforts. When the rather paranoid Pyral asked that Ickes guard his letters against those enemies who might use the information against him, Ickes did, although he hinted to Solicitor Nathan R. Margold that Pyral "probably overemphasizes the importance of the matter." See Dept. of the Interior, Office of the Secretary, Cent. Classified Files, 1907–1936, RG48, box 2000, files 12,13, pt. 6, NACP.

28. This analysis is based on a long letter from F. Emerson Hoar, consulting engineer, to Harold Ickes, February 17, 1936, in ibid., box 2001, files 12, 13, pt. 9, NACP. Hoar denied any particular interest in the matter and, by my reading, presented the secretary with a rational, objective view of the San Francisco situation.

29. Arthur Caylor, "The Whirligig," *San Francisco News*, n.d.

30. Memorandum for the Secretary, from Nathan R. Margold, Solicitor, n.d., Department of the Interior, Office of the Secretary, Cent. Classified Files, 1907–1936, RG48, box 2001, files 12, 13, pt. 9, NACP.

31. Memorandum for the Secretary, from Nathan R. Margold, Solicitor, December 15, 1936, in ibid., pt. 10, NACP. Also see Tim Redmond, "Hetch Hetchy Power Debacle," *San Francisco Bay Guardian*, 4, http://www.clovisnews.com/trails/hetch_hetchy_power.html.

32. Some of the chronology used comes from "Hetch Hetchy Bonds, 1941: Plan Nine With Comments," *San Francisco Call-Bulletin* Library, deposited in Vertical Files—Water—Hetch Hetchy, History Room, San Francisco Public Library. The documents and assembled chronology were undoubtedly for the use of the newspaper's reporters.

33. *U.S. v. City and County of San Francisco,* April 22, 1940, in 587 U.S. 7, 11 (1940).

34. The Rossi-Ickes telegram exchange may be found in "Hetch Hetchy Bonds, 1941: Plan Nine with Comments."

35. Harold L. Ickes, *The Secret Diary of Harold L. Ickes,* vol. 2, *The Inside Struggle, 1936–1939* (New York: Simon and Schuster, 1954), 422.

36. Figures from "Hetch Hetchy Bonds, 1941: Plan Nine with Comments."

37. See "Hearing Relative to the Hetch-Hetchy Power Project as Affecting the City of San Francisco by the Recent Ruling of the Supreme Court," Harold L. Ickes, Chairman, May 21, 1940, typescript in Dept. of the Interior, Office of the Secretary, Cent. Classified Files, 1937–1953, RG48, box 3870, file 12-47, pt. 4, NACP.

38. Elizabeth Cosby, Women's City Club of San Francisco, to Hon. Harold Ickes, May 30, 1939, in ibid.

39. See Donald J. Pisani, *Water and American Government: The Reclamation Bureau, National Water Policy, and the West, 1902–1935* (Berkeley and Los Angeles: University of California Press, 2002), 205–10.

40. *San Francisco Examiner,* August 2, 1940; Arthur Caylor, "Behind the News," *San Francisco News,* July 31, 1940.

41. U.S. Congress, House of Representatives, Committee on Public Lands, *Hearings on Amending Section 6 of the Act approved December 19, 1913 (38 Stat. 242), Commonly*

Known as the Raker Act 77th Cong., 2nd sess. (January 15–17, 19–24, 26, 27, 1942), H.R. 5964.

42. "Hearing Relative to the Hetch Hetchy Power Project as Affecting the City of San Francisco by the Recent Ruling of the Supreme Court," pp. 4, 5, 10, 11, 13.

43. Ibid., pp. 15, 16, 18, 96.

44. "Disposition of Hetch Hetchy Power by the City of San Francisco, California," typescript of meeting at the Department of Interior, August 21, 1940, 10–11, in ibid, pt. 7.

45. Hon. Harold L. Ickes to Mayor Angelo Rossi, December 4, 1940; Harold Ickes to Mayor Rossi, February 21, 1941; Rossi to Ickes, February 25, 1941; Ickes to Rossi, April 22, 1941; Ickes to Rossi, May 28, 1941, in "Hetch Hetchy Bonds, 1941, Plan Nine with Comments."

46. Redmond, "Hetch Hetchy Power Debacle," 7.

47. Memorandum for the Secretary from Abe Fortes, Division of Power, May 23, 1942, in Dept. of the Interior, Office of the Secretary, Cent. Classified Files, 1937–1953, RG48, box 3871, file 12-47, pt. 11, NACP. *San Francisco News*, April 28, 1942.

48. *San Francisco News,* July 22. 1943, December 1, 1944.

49. Excerpt from the Annual Message of Mayor Roger D. Lapham to the Board of Supervisors, January 2, 1945, in Dept. of the Interior, Office of the Secretary, Cent. Classified Files, 1937–1953, RG48, box 3872, file 12-47, pt. 12, NACP.

50. "Hetch Hetchy Development: General History," typescript for Dept. of the Interior use, n.d., in ibid., pt. 13.

51. For a detailed story of the development of the Turlock Irrigation District, see Alan M. Paterson, *Land, Water, and Power: A History of the Turlock Irrigation District, 1887–1987* (Glendale, Calif.: Arthur H. Clark Co., 1987). A less scholarly, yet helpful, history of the Modesto Irrigation District is Dwight H. Barnes, *The Greening of Paradise Valley: Where the Land Owns the Water and the Power* (Modesto, Calif.: Modesto Irrigation District, 1987).

52. See Barnes, *The Greening of Paradise Valley,* 94–104, passim.

53. See Oscar Chapman to Attorney General, September 1, 1950, and a file containing a number of SF power quarterly reports, both in Dept. of the Interior, Office of the Secretary, Cent. Classified Files, 1937–1953, RG48, box 3872, file 12-47, pt. 13, NACP. In his memo Chapman stated that after a review, he determined that "there is reasonable compliance . . . under the present arrangements, although the Department is likewise aware that San Francisco is not in strict compliance with its statutory provisions."

54. Most of the facts are taken from Warren D. Hansen, *A History of the Municipal Water Department and Hetch Hetchy System* (San Francisco: San Francisco Water and Power, 2002), 46–47.

55. Ibid. 48; Paterson, *Land, Water, and Power,* 306; interview with Patricia Martel by author, May 6, 2003.

56. Paterson, *Land, Water, and Power,* 378–79.

57. Ibid.; e-mail correspondence from Jerry Meral to the author, April 20, 2004.

9. THE LEGACIES OF HETCH HETCHY

1. Quoted in Wolfe, *Son of the Wilderness,* 342.

2. John Muir to Robert Underwood Johnson, January 1, 1914, RUJ Papers, C-B 385, "Muir" folder, 3, BL.

3. Jones, *John Muir and the Sierra Club,* 122.

4. Morrison, *J. Horace McFarland,* 173–93.

5. John Ise, in *Our National Parks,* 188, gives credit to Representative John Lacy of Iowa for first introducing a national park bureau bill in 1900. It went nowhere, as did attempts in 1902 and 1905.

6. Richard Sellars, *Preserving Nature in the National Parks: A History* (New Haven: Yale University Press, 1997), 40–41.

7. Ibid., 38–39.

8. Albright, Schenck, *Creating the National Park Service,* 125–26. Sellars points out that the plotters often met in Kent's Washington home "on the corner of F and Eighteenth streets—the same house where plans had earlier been formulated for passage of the Hetch Hetchy legislation." *Preserving Nature,* 302 n 46. Frederick Law Olmsted Jr., so eloquent in his defense of Hetch Hetchy and so instrumental in the molding of the "mission statement" of the Organic Act, had bowed out of the group, fearing conflict of interest with the Commission of Fine Arts, which was redesigning parts of Washington, D.C. See Morrison, *J. Horace McFarland,* 185.

9. Ise, *Our National Parks,* 190.

10. Albright, Schenck, *Creating the National Park Service,* 125.

11. Gifford Pinchot to Robert Underwood Johnson, April 17, 1905, RUJ Papers, C-B 385, "Pinchot" folder, BL.

12. Sellars, *Preserving Nature,* 36.

13. Gifford Pinchot to Marsden Manson, May 28, 1904, RUJ Papers, C-B 385, "Pinchot" folder, BL.

14. Sellars, *Preserving Nature,* 37. In the 1930s Pinchot would find himself and his Forest Service on the receiving end of an aggressive bureau when Interior Secretary Harold Ickes attempted to move the Forest Service into his proposed "department of conservation." As Pinchot was unsuccessful in swallowing the national parks, so was Ickes repulsed in his empire building as a result of what is often called the "Ickes-Pinchot brawl." See Miller, *Gifford Pinchot,* 341–56.

15. See John C. Miles, *Guardians of the Parks: A History of the National Parks and Conservation Association* (Washington, D.C.: Taylor and Francis, 1995), for a history of the association.

16. Mark W. T. Harvey, *A Symbol of Wilderness: Echo Park and the American Conservation Movement* (Albuquerque: University of New Mexico Press, 1994), 61.

17. Department of the Interior, "Report of the Director of the National Park Service," *Annual Report,* 1919, 962–63.

18. Quoted in Keith W. Olson, *Biography of a Progressive: Franklin K. Lane, 1864–1921* (Westport, Conn.: Greenwood Press, 1979), 107. See also Pisani, *Water and American Government,* 115.

19. *Congressional Record,* House, 66th Cong., 2nd sess., vol. April 19, 1920, 5856–57.

20. Donald C. Swain, *Wilderness Defender: Horace M. Albright and Conservation* (Chicago: University of Chicago Press, 1970) 122–23.

21. Albright, *Creating the National Park Service,* 333.

22. *Literary Digest* 67 (October 23. 1920), 90–91.

23. I agree with the suggestion in Karl Jacoby's work *Crimes against Nature: Squatters, Poachers, Thieves, and the Hidden History of American Conservation* (Berkeley and Los Angeles: University of California Press, 2001) that the setting aside of national parks and forests often had negative effects on the local populations. Such a theory, however, does not fit well with the Hetch Hetchy controversy, although there is a class-based argument that the defenders were insensitive to the poor working class of San Francisco.

24. "Another Hetch Hetchy," *The Outlook,* July 7, 1920, 448.

25. Michael J. Yochim, "Beauty and the Beet: The Dam Battles of Yellowstone National Park," *Montana The Magazine of Western History* 51 (Spring 2003), 26–27.

26. M. M. O'Shaughnessy to W. B. Lewis, November 14, 1919, in Central Files, Hetch Hetchy, 660-05-41, 1917–1922, YNPA.

27. Today certain areas of the Spring Valley watershed are opened to supervised walks.

28. Quoted in the *Sonora Banner,* June 1, 1928: also see *Stockton Daily Record,* July 10, 1928.

29. National Parks Service, *The National Parks: Index 1999–2001* (Washington, D.C., 2001), 30, lists Yosemite National Park at 761,266 acres. If 240,000 is accurate for the Tuolumne River watershed, it would be about one-third of the park. The photograph of O'Shaughnessy proudly points to "San Francisco's 420,000 acre water shed." However, this figure would include acreage in the Cherry Creek and Lake Eleanor area, which is in Stanislaus National Forest.

30. *Stockton Daily Record,* July 5, 1928.

31. Ibid., July 7, 1928.

32. Horace Albright to Stephen Mather, May 25, 1928, Central Files, Hetch Hetchy, 660-05.41, 1928–1930, YNPA.

33. *San Francisco Bulletin,* July 7, 1928.

34. SF City Engineers, 92/808 C, carton 3, folder "Site Inspection, May 1928," BL.

35. Robert M. Searles, San Francisco, to John Edwards, Assistant Secretary of the Interior, October 27, 1925; Stephen Mather to W. B. Lewis, November 3, 1925; Lewis to Mather, November 13, 1925, in Central Files, Hetch Hetchy, 660-05.41, 1928–1930, YNPA.

36. *San Francisco News,* June 21, 1928.

37. Quoted in ibid.

38. Press Release, Yosemite, October 5, 1927, Central Files, Hetch Hetchy (General), 601-04-4, 1926–1948, YNPA.

39. *Stockton Daily Record,* July 5, 1928.

40. Editorial, *Sonora Banner,* June 1, 1928.

41. Keith Trexler, "The Tioga Road," pamphlet (Yosemite Association, 1961), 23.

Trexler states that the Park Service built the roads with monies received from San Francisco as "rental" for the Hetch Hetchy area. Although the city did indeed make annual use payments to the park, it did so to fulfill the terms of the Raker Act.

42. "Second Supplement Agreement by the City and County of San Francisco, California under the Raker Act," as agreed on December 8, 1930, typescript in Central Files, City of San Francisco (Hetch Hetchy), 901-10, YNPA.

43. Roderick Nash, *Wilderness and the American Mind,* 4th ed. (New Haven: Yale University Press, 2001), 162. The first edition appeared in 1967, although Nash completed the dissertation version in 1964. His book arrived at a time when Americans were aware of, and many were enamored with, wilderness. It is my opinion that Nash's chapter on Hetch Hetchy was "thesis driven" in the sense that he transposed nature-related words into *wilderness*—which was not necessarily the intent of his primary sources.

44. Nash, *Wilderness and the American Mind,* 161–81; Clements, "Politics and the Parks," 214; Runte, *National Parks,* 78; Smith, *Pacific Visions,* 172; William Cronon, "The Trouble with Wilderness; or, Getting Back to the Wrong Nature," in William Cronon, ed., *Uncommon Ground: Rethinking the Human Place in Nature* (New York: W.W. Norton, 1995), 72.

45. Nash, *Wilderness and the American Mind,* 4th ed., 1.

46. Larry M. Dilsaver, ed., *America's National Park System: The Critical Documents* (Lanham, Md.: Rowman and Littlefield, 1994), 277–86.

47. For an extended discussion of the meaning of *wilderness,* see J. Baird Callicott, Michael P. Nelson, eds. *The Great New Wilderness Debate* (Athens: University of Georgia Press, 1998). I have been influenced by the views of historian Paul Hirt, particularly his comments in the session "The Challenge of Wilderness in the Changing American West," Western History Association Conference, San Diego, CA., October 4–7, 2001.

48. Marguerite S. Shaffer, *See America First: Tourism and National Identity, 1880–1940* (Washington, D.C.: Smithsonian Institute Press, 2001), 11; Marion Randall Parsons, "The Grand Canyon of the Tuolumne and the Merced," *Sierra Club Bulletin* 6 (January 1908), 6, 238; Lucy Washburn, "The Grand Circuit of the Yosemite National Park," *Sierra Club Bulletin* 7 (January 1910), 149–52.

49. Joseph L. Sax, *Mountains without Handrails: Reflections on the National Parks* (Ann Arbor: University of Michigan Press, 1980), 14.

50. "Articles of Incorporation," printed in Jones, *John Muir and the Sierra Club,* appendix B, 173. In 1951 the club's membership changed the purpose to "explore, enjoy, and preserve the Sierra Nevada and other scenic resources of the United States." See Cohen, *History of the Sierra Club,* 89–100, for a discussion of the change, which reflects a more contemporary view.

51. Jones, *John Muir and the Sierra Club,* 135. Jones quotes a large part of the Colby Brief in his book (135–45), but I have chosen to use a photocopy found in SCMP 71/295 C, Carton 38, folder 13, BL. Thus the national park quote may be found in Colby Brief, 7.

52. Colby Brief, 7–8.

53. Runte, *National Parks,* 19–20; William Owen, "The Matterhorn of America,"

Frank Leslie's Weekly, May 19, 1892; Allen Chamberlain, "Scenery as a National Asset," *Outlook* 95 (May 28, 1910), 162–64, 169.

54. Runte, *National Parks*, 89, suggests that the Colby Brief may have been a necessary compromise.

55. Chamberlain, "Scenery as a National Asset," 162–64.

56. Muir, *Our National Parks*, 3–4.

57. Quoted in Fox, *John Muir and His Legacy*, 59.

58. Roderick Nash, *The Rights of Nature: A History of Environmental Ethics* (Madison: University of Wisconsin Press, 1989), 40–41.

59. Muir, *The Mountains of California*, 278–91. Muir was a master of "recycling" his writings. To my knowledge the first time this passionate "call to arms" appeared was in the *Sierra Club Bulletin* 6 (January 1908), 211–20; also see Albanese, *Nature Religion in America*, 101–5.

60. See Marc Reisner, *Cadillac Desert: The American West and Its Disappearing Water* (New York: Viking, 1986), for a less-than-positive appraisal of the Bureau of Reclamation.

61. Harvey, *A Symbol of Wilderness*, 168.

62. Ibid., 57.

63. Ibid., 238–39.

64. David Brower, *For Earth's Sake: The Life and Times of David Brower* (Salt Lake City: Peregrine Smith Books, 1990), 343–44.

65. Stewart Udall, *The Quiet Crisis* (New York: Holt, Rinehart and Winston, 1963), 122.

66. See Byron E. Pearson, *Still the Wild River Runs: Congress, the Sierra Club, and the Fight to Save Grand Canyon* (Tucson: University of Arizona Press, 2002). Pearson gives Brower and the club credit but states that regional politics played an important role in the defeat of the proposal.

67. Monroe, *A Poet's Life*, 221.

10. RESTORATION

1. *Congressional Record*, 63rd Cong., 2nd sess., vol. 51 (December 5, 1913), 311.

2. "Interview with David Brower," May 27, 2000, at Hetch Hetchy Valley, by Ron Good, Chair of the Board of Directors, Restore Hetch Hetchy, p. 14, in Restore Hetch Hetchy Archives (abbreviated hereafter as RHHA).

3. Wallace Stegner, "Myths of the Western Dam," *Saturday Review*, October 23, 1965, 29–31.

4. A combination of environmental protest and regional political infighting defeated both the Marble Canyon and Bridge Canyon dam proposals. See Pearson, *Still the Wild River Runs.*

5. Christopher Swan, Chet Roaman, *YV88* (San Francisco: Sierra Club, 1977).

6. *San Jose Mercury News*, August 11, 1987; *San Francisco Examiner*, August 6, 1987, clippings in Vertical File Folder—Water, folder "Draining," San Francisco History Room, San Francisco Public Library.

7. *San Jose Mercury News*, August 11, 1987. James Watt seemed an anathema to en-

vironmentalists, but it was his propensity to make inappropriate decisions or comments that forced his resignation. When in September 1983 he publicly announced that on his staff "I have a black, a woman, two Jews, and a cripple," he sealed his fate.

His replacement, William P. Clark, had been Governor Reagan's chief of staff. A lawyer and later judge, Clark compiled a lackluster record as interior secretary.

8. *San Francisco Chronicle*, August 7, 1987.

9. Ibid.

10. *Sacramento Bee,* December 2, 1987.

11. Ron Stork, Friends of the River, to Bill Kahrl, *Sacramento Bee*, December 4, 1987; Chuck Washburn, Sierra Club, to Editor, *Sacramento Bee*, n.d., RHHA.

12. Charles Washburn, "No Hetch Hetchy or Auburn Dam," *Sacramento Bee*, September 20, 1987.

13. *Commonwealth Magazine,* January 29, 1988, 66.

14. Michael Fischer, "Notes on Conversation with Secretary Hodel at the Deadhorse Airport," August 12, 1987, in RHHA.

15. "Prepared Remarks of Secretary of the Interior Don Hodel to the FREEPAC," February 3, 1988, in RHHA; Brower, *For Earth's Sake,* 388.

16. *San Jose Mercury News*, September 9, 1987.

17. *Los Angeles Times,* September 9, 1987.

18. Reno *Gazette-Journal*, September 14, 1987; *San Jose Mercury News*, September 9, 1987.

19. "Visit to Hetch Hetchy Dam," memo to the Files by Michael McCloskey, Executive Director of the Sierra Club, September 9, 1987, RHHA.

20. "Meeting with Interior Secretary Hodel," memorandum by Mike McCloskey for the Hetch Hetchy Task Force, September 3, 1987, RHHA.

21. Ibid.

22. Memorandum from David Brower to Board of Directors, Sierra Club, September 24, 1987, in RHHA.

23. Included with Sierra Club comments in National Park Service, *Hetch Hetchy: A Survey of Water & Power Replacement Concepts* (February 1988), appendix.

24. Mayor Dianne Feinstein to Honorable Donald Paul Hodel, Secretary of the Interior, August 19, 1987, RHHA.

25. Ibid.; "Meeting with Interior Secretary Hodel," September 3, 1987.

26. National Park Service, *Hetch Hetchy: A Survey,* x–xi, xiii–xxiii.

27. Ibid., 10, 22.

28. Bureau of Reclamation, *Assessment '87: . . . A New Direction for the Bureau of Reclamation*" (1987).

29. National Park Service, *Hetch Hetchy: A Survey,* Appendix: Comments Received.

30. California State Legislature, Assembly Office of Research, *Restoring Hetch Hetchy* (1988), 26.

31. Ibid., 27, 40.

32. Ibid., 40–41.

33. George Perkins Marsh, *Man and Nature* (1864; Cambridge, Mass.: Harvard University Press, 1965), 35.

34. Frederick Turner, "The Invented Landscape," in A. Dwight Baldwin, Judith

De Luce, Carl Pletsch, eds., *Beyond Preservation: Restoring and Inventing Landscapes* (Minneapolis: University of Minnesota Press, 1994), 35.

35. Ibid., 36–37.

36. *New York Times,* October 15, 2002, A22.

37. Ken Brower, *Yosemite: An American Treasure* (Washington, D.C.: Special Publication Division, National Geographic Society, 1997).

38. John McPhee, "Encounters with the Archdruid," *The New Yorker,* April 3, 1971.

39. Ibid., 48.

40. Based on Adam Burke, "River of Dreams: The 30-Year Struggle to Resurrect Washington's Elwha River and One of Its Spectacular Salmon Runs," *High Country News,* 33 (September 24, 2001).

41. McPhee, "Farewell to the Nineteenth Century," *The New Yorker*, 75 (September 27, 1999), 47.

42. Burke, "River of Dreams," 11.

43. See Daniel McCool, "As Dams Fall, a Chance for Redemption," *High Country News* 36 (June 21, 2004), 12–13, 19.

44. McPhee, "Farewell to the Nineteenth Century," 50.

45. Most of the story and facts come from the author's telephone interview with Ron Good, September 27, 2003. For information on Gerold Meral, see Tim Palmer, *Stanislaus: The Struggle for a River* (Berkeley and Los Angeles: University of California Press, 1982), 60–68.

46. Information from fact sheet, October 20, 2002, http://www.sfwater.org. San Francisco Public Utilities District.

47. *San Francisco Chronicle*, November 6, 2002.

48. Mayor Dianne Feinstein to Jennifer Fosgate, August 21, 1987; Senator Dianne Feinstein to Ron Good, February 26, 2003, in RHHA.

49. Yosemite National Park, *Yosemite Guide,* vol. 30 (spring 2001). National Park Service.

50. *New York Times*, October 15, 2002, October 19, 2002; *San Jose Mercury News*, August 18, 2002.

51. *Los Angeles Times*, March 22, 2003.

52. For instance, see Sarah E. Null, "Re-assembling Hetch Hetchy: Water Supply Implications of Removing O'Shaughnessy Dam" (MA thesis, Department of Geography, University of California, Davis, 2003).

53. *Modesto Bee*, January 14, 2004.

54. Information from the Aspen Institute, http://www.aspeninstitute.org/Program. Accessed on April 9, 2003.

55. *Los Angeles Times*, February 5, 2004.

56. National Park Service, *Hetch Hetchy: A Survey,*" preface, iv.

AFTERWORD

1. Wallace Stegner, *The Sound of Mountain Water* (New York: Doubleday, 1969), 42.

2. Muir, *The Mountains of California 11*, 249.

3. William Frederic Badè, "Hetch Hetchy Valley and the Tuolumne Canyon," *Independent,* May 4, 1908, 1079–84, passim.

4. For some of these ideas I am beholden to Kendrick A. Clements, "Politics and the Park: San Francisco's Fight for Hetch Hetchy, 1908–1913," *Pacific Historical Review* 48 (May 1979), 214–15.

INDEX

Note: Italicized page numbers refer to illustrations